ibvt Schriftenreihe

Schriftenreihe des Instituts für Bioverfahrenstechnik
der Technischen Universität Braunschweig

Herausgegeben von Prof. Dr. habil. R. Krull

Band 73

Cuvillier-Verlag
Göttingen, Deutschland

Engineering a Biofilm

Imitating Physico-Chemical Properties
to Improve Mechanical Characterization

Von der Fakultät für Maschinenbau
der Technischen Universität Carolo-Wilhelmina zu Braunschweig

zur Erlangung der Würde
eines Doktor-Ingenieurs (Dr.-Ing.)
genehmigte Dissertation

von
Dipl.-Ing. Jan Hellriegel
aus Bergisch Gladbach

eingereicht am: 9. April 2014
mündliche Prüfung am: 13. Juni 2014

Gutachter: Prof. Dr. habil. R. Krull
Prof. Dr.-Ing. A. Kwade

2014

Insanity: doing the same thing over and over again and expecting different results.
Albert Einstein

To all my family and friends.

Danksagung - Acknowledgement

Zahlreiche Personen haben es mir ermöglicht, die nötigen experimentellen Daten zusammenzutragen und diese Arbeit zu verfassen. Ich danke

- meinem Doktorvater Prof. Dr. habil. Rainer Krull, der mir die Möglichkeit gegeben hat, im Rahmen des DFG-Projektes „Biofilmmechanik – Numerische und experimentelle Untersuchung der mechanischen Beanspruchung von Biofilmsystemen" zu promovieren, für die Diskussionen und Anregungen sowie die Freiheiten beim Anfertigen dieser Arbeit,

- Prof. Dr.-Ing. Arno Kwade für die Übernahme des Referats und die gute Kooperation mit dem Institut für Partikeltechnik,

- Prof. Dr.-Ing. Markus Böl für die Übernahme des Prüfungsvorsitzes sowie für die vielen ertragreichen Diskussionen und Anregungen zum mechanischen Verhalten von Hydrogelen und Biofilmen,

- Gena Peterat und René Stellmacher, meinen Bürokollegen, die mit mir gute und schlechte Tage durchgestanden haben, beide hatten stets ein offenes Ohr für meine Probleme und immer gute Ideen um weiter zu kommen,

- Steffi Günther für die langwierigen, oftmals schwierigen Messungen am Rheometer, die letztendlich zu guten Ergebnissen geführt haben,

- Antonio Bolea Albero für viele Stunden angeregter Diskussionen, in denen zahlreiche Ideen entstanden, verworfen und neue Erkenntnisse gewonnen wurden,

- allen technischen Mitarbeitern des Instituts für Bioverfahrenstechnik, Yvonne Göcke, Detlev Rasch, Sandra Hübner, Elena Kempf, Cord Hullmann und Rochus Jonas, ohne euch wäre ein effizientes Arbeiten nicht möglich gewesen,

- Frau Kahmann für den reibungslosen Ablauf in der Verwaltung,

- meinen studentischen Hilfskräften Lukas Schnöing und Yvonne Kraus für die Durchführung einiger Experimente,

- allen Mitarbeitern des Instituts für Bioverfahrenstechnik, es war eine sehr schöne Zeit und wir hatten viel Spaß zusammen,

- der Deutschen Forschungsgemeinschaft (DFG) (KR 1897/3-1) und der Max-Buchner-Forschungsstiftung der Dechema (Stipendium Nr. 2925) für die Finanzierung dieser Forschungsarbeit sowie

- meinen Studenten, die ihre Arbeiten zu Themen der Biofilmmechanik verfasst haben:

  - Florian Geppert (2012), Biofilmmechanik – Herstellung einer Transformante von *Pseudomonas putida* KT2440 zur Produktion von Grün Fluoreszierendem Protein (GFP)(Studienarbeit),

  - Daria Kaptsan (2012), Rheologische Charakterisierung und Etablierung von viskoelastischen Hydrogelen als Modellbiofilm (Masterarbeit),

  - Vincent Wiegmann (2012), Nutzung von Hydrogelen zur gezielten Biofilmbildung auf Oberflächen (Bachelorarbeit),

  - Tim Nadrowski (2013), Bestimmung von Sauerstoffdiffusionskoeffizienten in biofilmmimitierenden Hydrogelen (Bachelorarbeit),

  - Yvonne Kraus (2013), Einfluss der Medienzusammensetzung auf das Biofilmwachstum von *Pseudomonas putida* (Bachelorarbeit).

Jan Hellriegel   Braunschweig, im Juni 2014

Publications

Partial results of this work have been published in advance. This was authorized by the Department of Mechanical Engineering of the Technische Universität Braunschweig, Institute of Biochemical Engineering represented by Prof. Dr. habil. Rainer Krull.

Peer-reviewed Journals

Böl, M., Ehret, A. E., Bolea Albero, A., **Hellriegel, J.**, Krull, R. (2013) Recent advances in mechanical characterisation of biofilm and their significance for material modelling. Crit. Rev. Biotechnol. 33, 145-171.

Hellriegel J., Günther S., Kampen I., Bolea Albero A., Kwade A., Böl M., Krull R. (2014). A biomimetic gellan-based hydrogel as a physicochemical biofilm model. J. Biomater. Nanobiotechnol. 5, 83-97.

Conference contributions

Hellriegel, J., Bolea Albero, A., Böl, M., Krull, R. (2011). Hydrogele als biomimetische Modellsysteme zur Beschreibung des mechanischen Verhaltens von Biofilmen mit Hilfe von FEM. Vortrags- und Diskussionstagung: Bioverfahrenstechnik an Grenzflächen, Kurzfassungsband, P22, 103-104, Potsdam, Germany (Poster).

Hellriegel, J., Bolea Albero, A., Böl, M., Krull, R. (2011). Hydrogels imitating physico-chemical properties of biofilms to increase the accuracy of FEM based biofilm models. 1st European Congress on Applied Biotechnology, P39.05, Berlin, Germany (Poster).

Böl, M., Ehret, A.E., Bolea Albero A., **Hellriegel, J.**, Krull, R. (2012). Mechanical characterization of biofilm and their significance for material modeling. European Congress on Computational methods in Applied Sciences and Engineering, Micro-Symposium: Computational and experimental methods for mechanical analyses of microbial biofilms, MS 642-1, 2770, Vienna, Austria (Presentation).

Hellriegel, J., Krull, R. (2012). Applied online image acquisition techniques for surface visualization of biofilms. European Congress on Computational methods in Applied Sciences and Engineering, Micro-Symposium: Computational and experimental methods for mechanical analyses of microbial biofilms, MS 642-2, 3182, Vienna, Austria (Presentation).

Hellriegel J., Günther S., Kampen I., Kwade A., Krull R. (2014). Engineering a biofilm: A gellan hydrogel imitating physico-chemical properties to improve mechanical characterization. BRICS forum 2014, HZI Braunschweig, Germany (Poster).

Hellriegel J., Krull R. (2014). Engineering a synthectic biofilm based on a physico-chemical hydrogel model. Dechema Himmelfahrtstagung - Biomaterials - Made in Bioreactors, Proc., P17, Radebeul, Germany (Poster).

Abbreviations

a_i	polynomial coefficient	-
A_i	surface of biofilm	-
AB10	autoinducer bioassy 10	-
BDW	bio dry weight	gL^{-1}
BTR	biofilm tube reactor	-
BWW	bio wet weight	gL^{-1}
c	concentration	mM, gL^{-1}
CCD	central composite design	-
CSTR	continuous stirred tank reactor	-
CP	center point	-
CV	control volume	-
D	dilution rate	day^{-1}
D_F	diffusion coefficient	cm^2s^{-1}
D_{BF}	diffusion coefficient in biofilm	cm^2s^{-1}
DO	dissolved oxygen	mgL^{-1}, %
EPS	extracellular polymeric substances	-
EPS_{BF}	extracellular polymeric substances in the biofilm	-
EPS_{det}	detached extracellular polymeric substances	-
err(s)	error function	-
f	logarithmic frequency	s^{-1}
F	feed	Ld^{-1}
F_N	force	N
G	tensile or elastic modulus	Pa
G(t)	relaxation modulus	Pa
G_i	spring constant	Pa
G'	storage modulus	Pa
G''	loss modulus	Pa
G^*	complex shear modulus	Pa
GFP	green fluorescent protein	-
i	count, number of Maxwell element	-
I	intensity of LED	-
k	count, number of the hydrogel	-
LB	lysogeny broth	-
n	count, quantity of Maxwell elements	-

m	count	-
OD	Optical Density	-
q_{S_i}	net substrate flux	gm^{-2}
q_{X_i}	net biomass gain of biofilm	gL^{-1}
r	Bravais-Pearson correlation coefficient	-
R^2	squared Bravais-Pearson correlation coefficient	-
R	response	-
R'	power transformed response	-
RSM	response surface methodology	-
S	substrate concentration	gL^{-1}
S_{BF}	substrate in biofilm	gL^{-1}
S_{Bulk}	substrate in bulk	gL^{-1}
S_i	immobilized substrate concentration	gL^{-1}
SOC	super optimal broth with catabolite repression	-
T	temperature	$°C$
T_{SG}	sol-gel temperature	$°C$
U	voltage	-
V	volume	mL,L
V_i	volume of biofilm	mL,L
w	weight	g,kg
x_{CI}	confidence interval of x	-
$\bar{x}$	mean of x	-
x_{Gellan}	gellan content	% (w/v)
X	biomass concentration	gL^{-1}
X_{BF}	biomass concentration in biofilm	gL^{-1}
X_{Bulk}	biomass concentration in bulk	gL^{-1}
X_{det}	detached biomass concentration	gL^{-1}
X_i	immobilized biomass concentration	gL^{-1}
$Y_{X/S}$	yield coefficient	gg^{-1}

β	mass transfer coefficient	ms^{-1}
γ	strain	-
$\dot{\gamma}$	shear rate	s^{-1}
δ	phase angle	$^\circ$
η	dynamic viscosity	Pas
η^*	complex viscosity	Pas
$\varkappa$	decay rate	d^{-1}, h^{-1}
λ	power transformation factor	-
μ	growth rate	d^{-1}, h^{-1}
σ	stress	Pa
τ	relaxation time constant strength	s
ω	angular velocity, frequency	s^{-1}
$\vec{\omega}$	flow field	ms^{-1}

Contents

Abstract

Biofilms play a major role in material cycles and contribute to technical systems significantly. Despite their interference with the functionality of technical equipment or the product quality their ability to catabolize toxins and metabolize pharmaceutically relevant substances increases the interest in biofilm-based biotransformations. However, so far there is a lack of appropriate models that allow anticipating the mechanical stability of biofilms in particular during detachment processes.

The main objective of this work was the development of a hydrogel based physico-chemical and growth independent biofilm imitate to investigate mechanical, primarily fluid dynamical stresses and their influence on growth and detachment effects of biofilms. Verification was achieved by comparison with real single culture biofilms. Single culture biofilms of *Pseudomonas putida* KT2440 were cultivated in a biofilm tube reactor and grown on different surfaces, e.g., tube walls, surface-modified object slides, plastic and iron nettings as well as membrane filters. The establishment of on-line analytics allowed the automatic measurement of dissolved oxygen, pH, temperature and planktonic cell growth by optical density in the cultivation broth. Image acquisition of the biofilm surface supported the observation of biofilm development in terms of growth and detachment.

A hydrogel based on gellan was established as simplified artificial biofilm system, which behaves like a viscoelastic fluid. The degree of cross-linking at different gellan levels was modified by the addition of mono-and divalent ions (Na^+, Mg^{2+}) and the influence on the material constants in terms of storage (G') and loss (G'') modulus was determined. Experiments and evaluation in the predefined design space were supported by Central Composite Design (CCD), an experimental design technique. The resulting empirical model provides a suitable tool to mimic the mechanical behavior of real biofilms. The predicted and experimentally determined viscoelastic characteristic values of G' and G'' of the hydrogels deviated in the frequency range between 0.1 to 10 s^{-1} by less than 17.0 % and showed similar values as real single culture biofilms of *P. putida* KT2440.

The developed gellan-based hydrogel allows mimicking the mechanical properties of a biofilm excluding biological growth effects. It can now be used to validate further characterization methods or to test slowly growing biofilms where systematic errors are often smaller than the biological variances. Eventually this method enables a fast and reliable mechanical testing of biofilm systems.

Zusammenfassung

Biofilme spielen eine wichtige Rolle in Stoffkreisläufen und leisten einen bedeutenden Beitrag in technischen Systemen. Neben unerwünschten Biofilmen, die häufig eine Beeinträchtigung der Funktionsweise technischer Anlagen oder der Produktqualität verursachen, erhöhen ihre Fähigkeiten, Toxine zu katabolisieren und pharmazeutisch relevante Substanzen zu metabolisieren das allgemeine Interesse an der Biofilmforschung. Bislang fehlt es jedoch an geeigneten Modellen, mit denen sich die mechanische Stabilität von Biofilmen insbesondere bei Abtragsphänomenen prognostizieren lässt.

Das Hauptziel dieser Arbeit bestand in der Entwicklung eines auf Hydrogelen basierenden physikochemischen und wachstumsentkoppelten Biofilmimitats, um den Einfluss mechanischer, vorrangig fluiddynamischer Beanspruchungen auf die Integrität von Biofilmen genauer zu charakterisieren. Die Eignung des Biofilmimitats sollte durch einen Vergleich mit realen Reinkulturbiofilmen validiert werden.

Reinkulturbiofilme von *Pseudomonas putida* KT2440 wurden in einem Rohrreaktor auf unterschiedlichen Oberflächen, z. B. Rohrwänden, oberflächenmodifizierten Objektträgern, Kunststoff- und Drahtnetzen sowie Membranfiltern, kultiviert. Die Etablierung einer online-Analytik ermöglichte die automatische Erfassung des gelösten Sauerstoffs im Kultivierungsmedium, des pH-Wertes, der Temperatur und des planktonischen Zellwachstums über die Optische Dichte. Bildaufnahmen der Biofilmoberfläche dienten der Beobachtung der Biofilmentwicklung hinsichtlich Wachstum und Abtragsereignissen.

Als künstliches Biofilmsystem wurde ein Hydrogel auf Basis von Gellan etabliert, das sich ähnlich wie ein viskoelastisches Fluid verhält. Der Vernetzungsgrad wurde bei unterschiedlichen Gellan-Gehalten durch die Zugabe von ein- und zweiwertigen Ionen (Na^+, Mg^{2+}) variiert. Durch statistische Versuchsplanung mittels Central Composite Design (CCD) wurden mit Hilfe rheologischer Experimente in einem vorgegebenen Bilanzraum bei Variation des Gellan-Gehalts sowie der Na^+- und Mg^{2+}-Konzentrationen die Materialkonstanten, Speicher- (G') und Verlustmodule (G''), für die Biofilm-ähnlichen Hydrogele bestimmt. Das daraus resultierende empirische Modell stellt ein geeignetes Werkzeug dar, um das mechanische Verhalten realer Biofilme zu simulieren. Die prädiktierten und experimentell bestimmten viskoelastischen Kennwerte für G' und G'' des Hydrogels unterschieden sich im Frequenzbereich zwischen 0,1 und 10 s^{-1} um weniger als 17,0 % und zeigen ähnliche Werte auf wie reale Reinkulturbiofilme von *P. putida* KT2440.

Die entwickelte Methode erlaubt es, die mechanischen Eigenschaften eines Biofilms auf der Grundlage des Gellan-basierten Hydrogels ohne Berücksichtigung biologischer Wachstumseffekte zu bestimmen. Damit können nunmehr weitere Charakterisierungsmethoden getestet und Biofilme mit geringen spezifischen Wachstumsraten, deren Systemfehler oft kleiner als die biologische Varianz ist, überprüft werden. Letztendlich ermöglicht die Methode eine schnelle und zuverlässige mechanische Prüfung von Biofilmsystemen.

1 Introduction

1.1 Biofilms Today

Biofilm-forming microorganisms are ubiquitous and protected by a matrix of biopolymers, proteins and other organic substances [87]. In industrial processes these characteristics are responsible for a reduced plant productivity or poor product quality. Removal of these unwanted biofilms is still difficult. However, biofilms are able to degrade recalcitrant compounds and play an important role in the biological industrial wastewater treatment or as catalytic biofilms which are able to synthesize a wide variety of metabolites, potentially used to produce pharmaceuticals [26, 47]. Some of the metabolites are only available in well-defined mixed biofilm biocoenoses, increasing the interest in a controlled biofilm formation for their production [115]. By today researchers exploit biofilm-based biotransformations to produce biological toxic bulk chemicals, e.g., styrene oxide [57, 58].

Nevertheless, it is still difficult to predict biofilm growth and structural properties or cultivate biofilms with constant material characteristics. Limited knowledge of subpopulations, fluid-structure interactions of biofilm formation and detachment as a result of mechanical stresses do hitherto not allow reliable prediction and control of biofilm development. Recent studies suffer from a deficiency in experimental data to validate developed models [12, 13, 41].

1.2 Objectives

The main objective of this thesis is the development of a hydrogel based physico-chemical biofilm model to investigate fluid mechanical interaction and its influence on growth and detachment effects of biofilms. Therefore, a representative and reproducible biofilm is required. Thus a microbial test system needs to be established which allows further analysis of planktonic and sessile growth in a controlled environment.

A further objective is to obtain a mono-septic culture biofilm to reduce the complexity of a multispecies biocoenosis. Targeted surfaces for biofilm development are either the walls of the tubing, surface modified object slides, plastic or iron nettings and membrane filters. As far as possible automated on-line analytic should be used to measure dissolved oxygen, pH, surface and planktonic cell growth during cultivation. A suitable image acquisition technique to measure and visualize surface attached biofilm development in terms of growth and detachment processes is required.

Established measurement techniques are validated with real biofilms and with the help of the established hydrogel model. For future visualization of cell distribution within the biofilm and hydrogels with a confocal laser scanning microscopy (CLSM) the model strain *P. putida* KT2440 is genetically modified to produce the green fluorescent protein (GFP).

The development of a physico-chemical biofilm model based on highly hydrolyzed gellan hydrogels is supported by experimental design techniques. The resulting gellan-based hydrogel should be validated to be a mechanically similar biofilm imitate. Finally, the predicted model system should withstand a comparison with real biofilms and a physical biofilm model based on worm-like-chains. Its purpose is to mimic the viscoelastic properties of biofilms in terms of storage (G') and loss (G'') moduli. The physico-chemical model is supposed to imitate biofilms of different strength by changing its composition resulting in biofilm like viscoelastic behavior.

2 Theory

2.1 General Knowledge of Biofilms

Biofilms are considered to be a community of microorganisms attached to surfaces or phase boundaries such as solid-liquid, solid-gaseous or gaseous-liquid [28, 34]. They have been around for two to three billion years and grow ubiquitously on all terrestrial surfaces. The first discovery of microbial life is linked to the breakthrough of the microscopy initiated by Robert Hooke (1635 - 1703), who first illustrated fruiting structure of molds and published it in 1665. Eleven years later Antoni van Leeuwenhoek used a simple microscopic construction to first visualize bacteria [95]. It was not until the nineteenth century that research in microbiology became of general interest [95] and not before the mid twentieth century that surface attached microorganisms were found to play a major role [16], even though it is the most common mode of growth for microorganisms on earth [28, 151, 155]. The production and secretion of exopolymers allow bacteria to modify the hydrophobicity of surfaces and assist adsorption and colonization of any substratum [111]. Some intensively studied bacteria are *Pseudomonas aeruginosa* or *Streptococcus mutants* which as pathogenic microorganisms are responsible for a wide range of infections. They produce bacterial alginates or biopolymers that form a hydrogel-like environment as protection matrix. Most biofilms are heterogeneous and are characterized by a nutrient gradient from the surface to the substrate as a result of limited diffusion through the exopolymeric matrix. The matrix characterized by extracellular polymeric substances (EPS) also provides mechanical stability and functions as reservoir for nutrients or water [111].

2.1.1 Structure and Composition of Biofilms

According to Mayer et al. [98] biofilms are highly hydrated viscoelastic systems with a water content between 92 to 94 % and a high storage (G') but a small loss (G'') modulus (compare Chapter 2.3.1). They found a bacterial concentration of 5 to $7 \cdot 10^{11}$ cells·g_{BDW}^{-1}. Costerton et al. [34] reported a biomass content for dried biofilms between 10 to 25 % and 75 to 90 % EPS, respectively. Biofilm structures and properties are influenced by medium composition and environmental conditions.

Fig. 2.1 shows a biofilm model adapted from Möhle [103]. The rough biofilm surface topology facilitates an efficient nutrient uptake and allows a constant fluid flow ($\vec{\omega}$) through the channels [134]. Fluid induced stresses (σ) into the biofilm cause detachment by erosion and sloughing (X_{det}) which eventually leads to stronger more compact biofilms and an increased strength of the EPS-matrix [3]. In laminar flows, biofilms tend to be fluffy and in general thicker than in tubular flow, which results in thinner and denser biofilms and promotes the formation of streamers [104, 148, 157]. Slow flowing fluids promote the biofilm to build mushroom-like structures. The controlling instance of biofilm formation is still poorly understood.

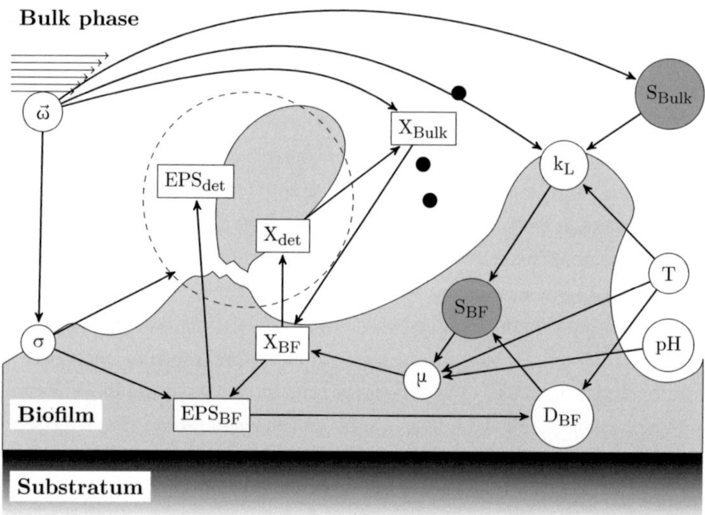

Figure 2.1: Biofilm model, adapted from Möhle [103] and modified. Fluid flows over the biofilm surface ($\vec{\omega}$), induces stress (σ) into the biofilm and transports biomass (X_{Bulk}) and substrate (S_{Bulk}). The stress influences the EPS structure (EPS_{BF}) and causes erosion or detachment (EPS, X_{det}). Planktonic biomass (X_{Bulk}) (re)attaches to the substratum, increasing the biomass of the biofilm (X_{BF}). Substrate (S_{Bulk}) is transported over the boundary layer (k_L) and diffuses (D_{BF}) through the biofilm. The specific growth rate (μ) is influenced by the substrate availability (S_{BF}), temperature (T) and pH (pH) in the biofilm and related the production of biomass (X_{BF}) as well as the EPS (EPS_{BF}).

Wäsche [172] related the biofilm topology to the substrate availability and stress intensity. High substrate concentration at moderate flow velocities resulted in increased biofilm growth but in a weak and fluffy EPS-matrix. The author confirmed that at high stresses the fluid-biofilm interaction exceeds the influences of substrate and determines the biofilm surface structure and thickness.

Tolker-Nielsen et al. [165] investigated the development and dynamics of *Pseudomonas* sp. biofilms with confocal laser scanning microscopy (CLSM). Their published data proposes that the initial biofilm attachment is caused by simple growth on the substratum. They also identified fast movement of *P. putida* OUS82 bacteria inside microcolonies and discussed the possibility of bacteria being in different physiological states when sessile or planktonic. However, growing on surfaces *Pseudomonas* sp. strain B13 bacteria formed ball-shaped microcolonies and *P. putida* OUS82 showed loose protruding structures, displaying different kinds of biofilm formation. Nancharaiah et al. [109, 110] quantitatively described a *P. putida* KT2440 biofilm in situ and used GFP, a red fluorescent protein (DsRed) and a nucleic acid stain (SYTO 60) to study conjugal gene transfer. Their flow cell experiments showed a percentage surface coverage of 80 % in 48 hours. This proved the possibility to monitor biofilms with a CLSM and showed a significant correlation between the GFP and SYTO 60 stain distribution.

Nilsson et al. [116] described two gene clusters, pea and peb, in *P. putida* KT2440 to stabilize the biofilm. They postulated that biofilm development might occur without exopolysaccharides since these only function as stabilizers. In contrary, Klausen et al. [77] saw no reason of biofilm development of the model biofilms *P. aeruginosa* and *P. putida* in genetic program, but in a number of different mechanisms to colonize surfaces. The nutrient availability in a biofilm is defined by the growth limiting substrate concentration in the bulk phase (S_{Bulk}), the mass transport through the boundary layer (k_L) and the diffusion coefficient within the biofim (D_{BF})[172]. Biofilm growth (μ) is a function of substrate concentration (S_{BF}), pH and temperature (T) and closely linked to the production of EPS (EPS_{BF}).

Certain environmental conditions, e.g., ion and substrate concentrations, pH, humidity or toxicity, can promote or reduce biofilm growth and strengthen or weaken the EPS-matrix [88]. EPS production is often increased by temperatures lower than optimal and high carbon to nitrogen ratios as well as a lack of nutrients. Chang et al. [25] published data showing that for the bacteria *P. putida* mt-2 the production of alginate is linked to the availability of water functioning as reservoir.

2.1.2 Composition of the Extracellular Polymeric Substances

According to Flemming et al. [48] most of the carbohydrate fraction of the EPS are polysaccharides and have been intensively studied since their commercial application as gelling agent, flocculates, foam stabilizer, hydrating agents and biosurfactants. Their gel-like behavior stabilizes and gives structure to the biofilm. Literature differentiates between non-bacterial and bacterial polysaccharides because they have dissimilar molecular weights and properties. It is, therefore, only possible to compare data qualitatively [87]. The average molecular length ranges from 500 to 2000 kDa. Bacterial polysaccharides can be capsular or released to the surrounding. Capsular polysaccharides are believed to be connected to the cell surface, possibly even covalently bonded while the released ones are free [1, 83, 154]. There is a wide variety of bacterial polysaccharides, some of them are homo-polymers like dextran, cellulose or sialic acid. However, most of them consist of multiple mono sugars, e.g., emulsan, xanthan, alginate or gellan [153].

Celik et al. [19] cultivated *P. aeruginosa* G1 and *P. putida* G12 on glucose, mannose, fructose and xylose and characterized the EPS of these strains. The monosaccharide composition of the biofilms can be analyzed by HPLC. Celik et al. used a PAP medium with glycerol, 2 % glucose (w/v), 2 to 6 % mannose, fructose or xylose (w/v) as carbon source for EPS production and found a high content of neutral sugars (92.0 to 99.2 %) and acetylated amino sugars (0.8 to 8.0 %) in the EPS. Kachlany et. al. [72] investigated the structure and carbohydrates involved in the exopolysaccharide capsule of *P. putida* G7. Special for *P. putida* G7 is the ability to produce capsular exopolymers which surround the young bacteria cells in planktonic and sessile state, respectively. Older cells lost the capsular EPS, thus resulting in cell associated and non-associated EPS. The EPS was purified and analyzed. It contained the monosaccharides, glucose, rhamnose, ribose, N-acetylgalactosamine and glucuronic acid. Jahn et al. [69] described the composition of *P. putida* biofilms. They had a maximum specific growth rate of $\mu_{max} = 0.34$ h^{-1} on 10 gl^{-1} citrate and of $\mu_{max} = 0.28$ h^{-1} on 10 gl^{-1} glucose in batch at 30 °C. The composition of the biofilm in a continuous cultivation varied depending on the limitation of either the carbon source or oxygen supply. The authors found high concentrations of proteins, polysaccharides, uronic acid and DNA. It is common that one bacterial strain produces multiple polysaccharides [154], thus the EPS always is a group of different chemical molecules.

2.1.3 Biofilm Mechanics and Extracellular Polymeric Substances

Determining the mechanical properties of real biofilms or EPS-matrix is still difficult. The produced polysaccharide and the amount depend on the availability of certain substrates and nutrients. The nitrogen or phosphorus ratio to carbon can determine if a cell uses energy for growth or EPS production. A high carbon concentration usually promotes EPS production [153]. Mayer et al. [98] used *P. aerigonosa* SG81 biofilms grown on agar in petri dishes for 24 h at 36 °C to prepare a bacterial EPS-solution. To separate the biomass from its matrix they carefully removed the bacterial lawn from the agar surface, suspended and homogenized it by vigorous stirring in sterile water. Extraction was achieved by centrifugation and filtration, finally resulting in an EPS-solution. The authors found three binding forces, London dispersion forces, electrostatic interactions and hydrogen bonds, between the EPS molecules. Their strength depended on the arrangement of the monomers, linear or branched, and their composition and conformation. High uronic acid content usually stabilizes the hydrogel while a high arabinose content induces cell aggregation [98].

Körstgens et al. [81] described the mechanical stability of biofilms upon uniaxial compression. They stated that the mechanical properties of the EPS-matrix are mostly defined by the polysaccharidic skeleton. Körstgens [79] and Wloka [178] proceeded similar to Mayer et al. [98] to separate the EPS from the mucoid bacterial lawn and rheologically characterized an EPS-solution. However, the procedure required scraping off the biofilm and this destroyed the original matrix of the EPS. The characterized EPS-solution had little in common with the EPS-matrix of a biofilm. The mechanical data for the cohesive strength, storage, loss, elastic or shear moduli reported in literature varies over several magnitudes [12]. Reasons are a wide range of measurement methods, such as centrifugation or tension devices, fluid shear techniques, erosion or compression experiments and rheometry.

Furthermore, alternating species of biofilms or different culture strategies change the general structure of the biofilm. Möhle et al. [104] reported a cohesive strength of 6.0 to 7.7 Pa for a mixed culture biofilm while Poppele et al. [133] found up to $1.6 \cdot 10^4$ Pa for *P. aeruginosa* biofilms. The elastic modulus for *P. aeruginosa* biofilm was given with 1.0 to 5.0 Pa by Klapper et al. [76] or with up $2.3 \cdot 10^4$ Pa for *Staphylococcus epidermidis* by Hohne et al. [60]. Values for the storage (G') and loss (G'') moduli reported by Characklis [27] for a mixed biofilm are 60 Pa and 120 Pa, respectively, while Moresi et al. [105] found them to be $4.2 \cdot 10^4$ and 400 Pa for the EPS gel extracted from *Azotobacter vinelandii* biofilms.

Fig. 2.2 shows some sample data for G' and G'' for a gellan based hydrogel, a biofilm of *S. aureus* 6ME and the EPS-solution. The hydrogel contained $x_{Gellan} = 0.7\,\%(w/v)$ and $c_{Ca^{2+}} = 0.03\,\%(w/v)$, probe hight was given with 5.5 mm, $\gamma \approx 1.8\,\%$ and data was extracted from Oliveira et al. [127]. The *S. aureous* biofilm was stressed with 3 Pa and data taken from Di Stefano et al. [37]. Wloka et al.'s [179] EPS-solution was measured after addition of 7 mM Ca^{2+} at a strain of $\gamma \approx 10\,\%$. The values for G' and G'' are in the same magnitude for the hydrogel and the biofilm, with approximately 10 to 40 kPa for G' and 2 to 30 kPa for G'', respectively. The reported values of the EPS-solution are several magnitudes below the values for the biofilm with $20 \cdot 10^{-3}$ for G' and $20 \cdot 10^{-4}$ for G''. Thus, the analogy of the EPS-solution to the biofilm is questionable, whereas the hydrogel and the biofilm clearly show similarities. A comprehensive summary of mechanical methods and experimental determined data on biofilms can be found in the review by Böl et al. [12].

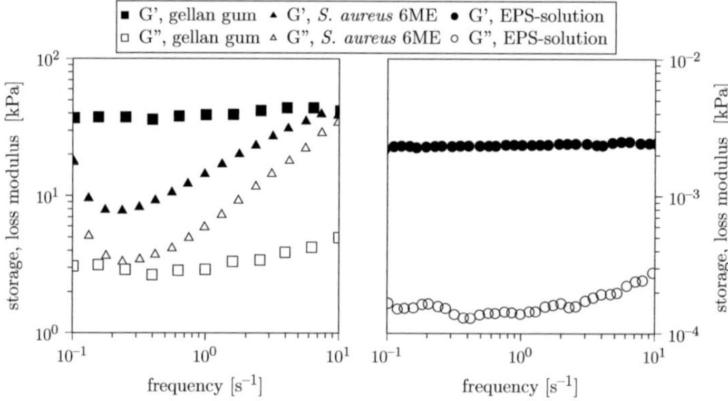

Figure 2.2: Storage (G') and loss (G'') moduli for a hydrogel (squares), a biofilm of *S. aureus* 6ME (triangle) and an EPS-solution (circle). The hydrogel contained $x_{Gellan} = 0.7\,\%(w/v)$ and $c_{Ca^{2+}} = 0.03\,\%(w/v)$, probe hight was 5.5 mm, $\gamma \approx 1.8\,\%$. Data was extracted from Oliveira et al. [127]. The *S. aureus* 6ME biofilm was stressed with 3 Pa. Data was extracted from Di Stefano et al. [37]. The EPS-solution was measured after addition of 7 mM Ca^{2+} at a strain of $\gamma \approx 10\,\%$. Data extracted from Wloka et al. [179].

2.1.4 Biofilm Detachment

The biofilm life cycle is a balance between microbial growth and detachment processes [163]. There are multiple reasons besides flow induced stresses [132] that promote detachment effects such as quorum sensing [137], nutrient completion [62] or the growth history [163]. Earlier events such as substrate limitations or changes in hydrodynamic forces have influenced the structure during the biofilm development [65]. Mushroom-like shaped biofilms are more easily destroyed by an increase in flow induced stresses than smooth surfaces. The cohesive strength of the biofilm is a result of the EPS-matrix and as such contingent on the availability of nutrient [75, 172].

Detachment is usually categorized into two types, sloughing and erosion [132]. The first is described as the tearing off of large biomass fragments, resulting in rougher surfaces or even complete removal of the biofilm [169]. The latter is the removal of smaller biofilm fragments which smoothen the biofilm surface [163]. Fast grown biofilms show a higher detachment rate than slowly grown biofilms if stressed similarly since high growth rates cause a weaker EPS-matrix and thus unstable biofilm accumulation [146]. Erosion and sloughing also provide biomass for re-attachment or population of other surfaces further down the stream [135].

2.1.5 Influence of Metal Ions

Körstgens [79] described a decreasing stability in the order of these metal ions: Cu^{2+} > Al^{3+}, Ca^{2+}, Fe^{3+} > Mn^{2+}, Fe^{2+} and Co^{2+}. The concentration of these ions play a major role for the biofilm strength and often have an optimal concentration after which no change or a decrease in stability was monitored. The author explained the strengthening effect of the ions by an increased electro static interaction between the multivalent cations and the extracellular polysaccharides, e.g., alginate. Wloka [178] analyzed the influence of Ca^{2+}, Mg^{2+} and Mn^{3+} on the storage and loss moduli of *P. aeruginosa* biofilms. He described entanglements and Coulomb forces as major reasons for the cross-linking of the EPS-solution and the biofilms with a decreasing influence of Ca^{2+}, Mn^{3+} and Mg^{2+}. An increase in the ion concentration changed the dominating attraction from entanglements to electro static forces.

Fig. 2.3 shows the results from Wloka et al. [180] for *P. aeruginosa* biofims with 1 or 10 mM Ca^{2+} in comparison to data published by Lieleg et al. [88] who investigated the influence of Ca^{2+}, Cu^{2+} and Fe^{3+} on the biofilm elasticity. There is an obvious increase in the biofilm storage and loss moduli due to the addition of Ca^{2+}

concerning Wloka's data. However, the data from Lieleg et al. [88], who cultured *P. aeruginosa* on LB medium, is within the same range. But LB medium is known for its low Ca^{2+} and Mg^{2+} concentration [175]. This shows that the comparability of the data between different authors and experiments is difficult.

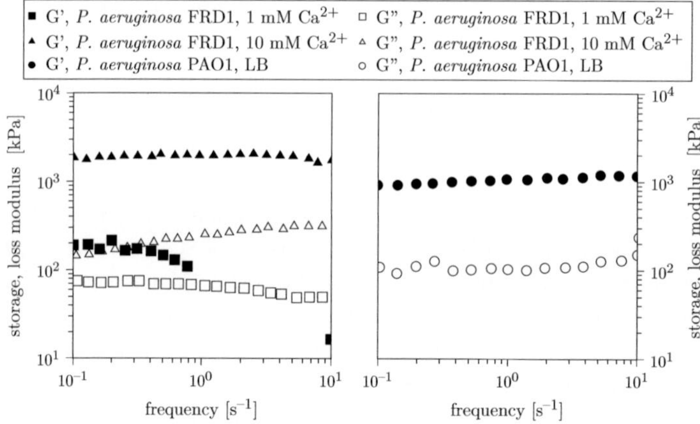

Figure 2.3: Storage (G') and loss (G'') moduli of *P. aeruginosa* FRD1 and PAO1 biofilms. They were grown with 1 and 10 mM Ca^{2+} or in LB medium. Data extracted from Wloka et al. [180] (squares, 1 mM Ca^{2+}, triangles, 10 mM Ca^{2+}) and Lieleg et al. [88] (LB, circles).

2.1.6 Basics on Batch, Continuous and Biofilm Growth Models

Batch growth is characterized by a lag-, growth-, stationary- and death-phase. A simple mathematic description for bacteria growth is based on the division of a cell after some time. Instead of cell numbers it is more common to use the dry cell weight or bio dry weight per volume, a concentration (BDW, X) [11]. For unlimited growth, the change in planktonic biomass solely depends on the total amount of biomass. This can be written as

$$\frac{dX}{dt} = \mu X \tag{2.1}$$

with μ being the specific growth rate [11]. After integration and simplification of Eq. (2.1) with respect to time the biomass concentration is

$$X = X_0 e^{\mu(t-t_0)} \tag{2.2}$$

where X_0 is the biomass concentration at t_0 which is the start time of the exponential growth phase. The decline during the death phase can be described analogous to the exponential growth Eqs. (2.1) and (2.2) as

$$\frac{dX}{dt} = -kX \tag{2.3}$$

and

$$X = X_x e^{-\varkappa(t-t_x)} \tag{2.4}$$

where X_x represents the biomass concentration at the end of the stationary phase t_x. A simple relation between growth and substrate concentration is given by the Monod model [11]

$$\mu = \mu_{max}\frac{S}{K_S + S} \tag{2.5}$$

with μ_{max} being the maximum specific growth rate and K_S the Monod or half-velocity constant, with $\mu = 0.5\,\mu_{max}$ for $S = K_S$. The yield coefficient $Y_{X/S}$ is defined as the mass of cells produced per mass of substrate utilized

$$Y_{X/S} = \frac{dX}{dS}\,. \tag{2.6}$$

Fig. 2.4 presents a scheme of the simplified test site (CV test site) which has been divided into two sections, the feed/gassing chamber (CV feed) and the biofilm tube reactor (BTR) section (CV BTR). The feed/gassing chamber is connected to the biofilm tube reactors and has two inlets with a feed rate of F_{in} and F_2, two outlets with a flow rate of F_{out} and F_2. The volume is given as V_1. Consequently, the BTR section has a feed F_2 and an identical outlet flow rate with a volume of V_2. Biofilm build-up is ideally limited to the BTR section, introducing biomass and substrate source or sink terms (q_{X_i}, q_{S_i}) with a biofilm specific growth rate of μ_i. The biofilm volume is V_i and the surface area A_i with an immobilized biomass of X_i and a substrate concentration of S_i. The complete test site has a single feed F_{in} with a substrate and biomass concentration of S_0 and X_0 and an outlet with a flow rate of F_{out} with a substrate and biomass concentration of S and X, respectively.

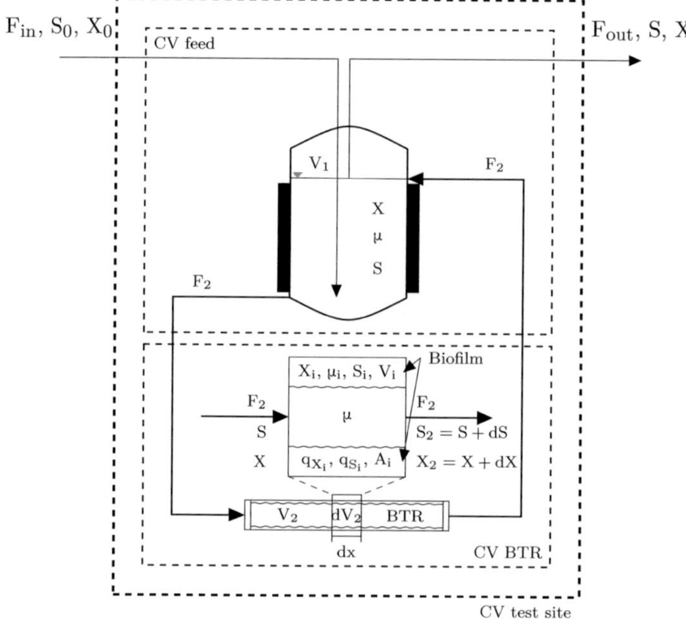

Figure 2.4: Schematic presentation of the microbial test site (CV test site) divided into two sections, the feed/gassing chamber (CV feed) and the biofilm tube reactor (BTR) section (CV BTR).

The total volume of the test site is $V = V_1 + V_2 + V_i$. The volume of the complete system is kept constant

$$\frac{dV}{dt} = F_{in} - F_{out} = 0 \rightarrow F_{in} = F_{out} = F \tag{2.7}$$

and F_{in} contains no biomass $X_0 = 0$. The mass balances for biomass and substrate for CV feed are

$$V_1 \frac{dX}{dt} = -FX - F_2 X + F_2 X_2 + \mu X V_1 \tag{2.8}$$

and

$$V_1 \frac{dS}{dt} = FS_0 - FS - F_2 S_2 - \frac{1}{Y_{X/S}} \mu X V_1, \tag{2.9}$$

where the volume V_1 is constant with respect to time. Setting up the mass balances for biomass and substrate for the volume section dV_2 of the BTR (CV BTR), assuming that the volume of the biofilm (V_i) is negligible small compared to the volume of the BTR (V_2),

$$dV_2 - dV_i \approx dV_2 \tag{2.10}$$

results in

$$dV_2 \frac{dX}{dt} = F_2 X - F_2 (X + dX) + \mu X dV_2 - q_{X_i} dV_i \tag{2.11}$$

and

$$dV_2 \frac{dS}{dt} = F_2 S - F_2 (S_2 + dS_2) - \frac{1}{Y_{X/S}} \mu X dV_2 - q_{S_i} dA_i. \tag{2.12}$$

The changes of biomass and substrate within the volume element dV_2 are the sums of the biomass and substrate, entering, leaving, growing, attaching and detaching or being consumed. Biofilm development is included by the biomass and substrate source or sink terms, q_{X_i} and q_{S_i}, which describe the transport of biomass per volume of biofilm to the biofilm (attachment and detachment) and the diffusion of substrate through the boundary layer into the biofilm. q_{X_i} and q_{S_i} are time dependent functions of X_i, S_i and μ_i as discussed in Chapter 2.1.1 and visualized in Fig. 2.1.

However, with respect to the whole system the amount of biofilm is small and in terms of the planktonic growth biofilm development slow, thus can be considered as quasi stationary. The flow rate F_2 is set to 1.5 $Lmin^{-1}$ (compare Chapter 3.1.5) and the volume of V_2 is approximated with 0.6 L resulting in a residence time of $F_2/V_2 = 24$ s. Considering a maximal specific growth rate of $\mu = 0.59$ h^{-1} the increase in biomass and the consumption of substrate in 24 s can be neglected, which results in

$$dS|_{t=24\text{ s}} = 0 \rightarrow S_2 = S \tag{2.13}$$

and

$$dX|_{t=24\text{ s}} = 0 \rightarrow X_2 = X. \tag{2.14}$$

Eqs. (2.11) and (2.12) can be simplified to

$$V_2 \frac{dX}{dt} = \mu X V_2 - q_{X_i} V_i, \tag{2.15}$$

$$V_2 \frac{dS}{dt} = -\frac{1}{Y_{X/S}} \mu X V_2 - q_{S_i} A_i \tag{2.16}$$

and Eqs. (2.8) and (2.9) to

$$V_1 \frac{dX}{dt} = -FX + \mu X V_1, \tag{2.17}$$

$$V_1 \frac{dS}{dt} = F(S_0 - S) - \frac{1}{Y_{X/S}} \mu X V_1. \tag{2.18}$$

The total volume is given as

$$V = V_1 + V_2 \tag{2.19}$$

and the total changes in biomass and substrate are

$$V \frac{dX}{dt} = V_1 \frac{dX}{dt} + V_2 \frac{dX}{dt} \tag{2.20}$$

$$V \frac{dS}{dt} = V_1 \frac{dS}{dt} + V_2 \frac{dS}{dt} \tag{2.21}$$

Combining Eqs. (2.15), (2.17) and (2.20) and Eqs. (2.16), (2.18) and (2.21), respectively yields

$$\frac{dX}{dt} = -\frac{F}{V} X + \mu X - q_{X_i} \frac{V_i}{V} \tag{2.22}$$

and

$$\frac{dS}{dt} = \frac{F}{V}(S_0 - S) - \frac{1}{Y_{X/S}} \mu X - q_{S_i} \frac{A_i}{V}. \tag{2.23}$$

The volume of the immobilized biomass (V_i) compared to the test site volume (V) is very small and can be ignored if interest lies in balancing the whole system. Eqs. (2.22) and (2.23) are further simplified and equal the mass balances of a continuous cultivation in chemostat or an ideal continuous stirred tank reactor (CSTR).

Conditions are constant with respect to time [11]:

$$0 = \frac{F}{V}\left(X_0 - X\right) + \mu X, \tag{2.24}$$

$$0 = \frac{F}{V}\left(S_0 - S\right) - \frac{1}{Y_{X/S}}\mu X. \tag{2.25}$$

The dilution rate D is defined as the ratio of F/V [11]. Since the system is at steady state

$$\frac{dX}{dt} = \frac{dS}{dt} = 0 \tag{2.26}$$

Eq. (2.24) is simplified to

$$DX = \left(D - \mu\right) X. \tag{2.27}$$

Eq. (2.27)'s only solutions are X $=$ 0 and D $=$ μ [11]. Substituting D for μ in Eq. (2.5) and solving for S yields

$$S = \frac{DK_s}{\mu_{max} - D} \quad \text{for } X > 0. \tag{2.28}$$

Inserting Eq. (2.28) into Eq. (2.25) the biomass concentration for a CSTR is given as

$$X = Y_{X/S}\left(S_0 - \frac{DK_s}{\mu_{max} - D}\right). \tag{2.29}$$

If $D > \mu_{max}$ the system becomes unstable and more biomass leaves the system than is produced. Thus the biomass concentration drops until it reaches a new steady state [11]. This effect is called wash out.

Eq. (2.22) includes the biomass change q_{X_i} of the biofilm through attachment or removal processes as result of erosion, sloughing or growth. It is a function of the nutrient concentration, the specific growth rate μ and e.g., pH within the biofilm. The biofilm development is not constant with respect to time. q_{S_i} is the net transport of nutrient into the biofilm which depends on the mass transfer across the phase boundary and is limited by diffusion. The resulting Eqs. (2.22) and (2.23) are time dependent and not easily solvable thus numerical models have been developed.

There are multiple biofilm models to describe the biofilm development and growth. One of the first 1-D models was published by Wanner et al. [170] and was able to compute concentration gradients perpendicular to the biofilm-liquid interface. Later 2D and 3D individual-based models were developed by Wimpenny et al. [177], Picioreanu et al. [130–132], van Loosdrecht et al. [168] and Chambless et al. [22].

Some fluid based, viscoelastic or elastic models were presented by Dockery et al. [38], Alpkvist et al. [3–5], Towler et al. [166], Taherzadeh et al. [157], Coroneo et al. [32] and Bolea Albero et al. [14]. A promising model was published by Ehret et al. [41] who used a network theory to model biofilms as a superposition of worm-like-chains. Nevertheless, all models lack some basic understanding of the real biofilm mechanics and depend on many assumptions [12].

2.1.7 Physical Biofilm Models

Besides the discussed biofilm computer models some physical models have been developed. Rowley et al. [138] used alginate hydrogels as EPS materials for mammalian cells. They modified the alginate and covalently bonded cell ligands to grow mouse skeletal myoblasts but were not interested in a mechanical model imitate. Strathmann et al. [152] used homogeneous and porous agarose beads with immobilized cells to simulate the EPS-matrix and studied biofilm population distribution and transport processes. They successfully simulated the behavior observed in thick highly stressed biofilms with their typical substrate gradients, diffusion limitation and biomass distribution. Elasri et al. [44] and Wloka et al. [179] used an alginate matrix to simulate biofilms. While Elasri et al. [44] studied the response to UV radiation and found that the damage on cells was limited by the protective property against UV light of the hydrogel, Wloka et al. [179] used alginate for rheology measurements and showed, probably firstly, the mechanical similarity of hydrogels and biofilms.

2.2 Hydrogels

A gel is a cross-linked polymer network, the space of which has been filled with a liquid, e.g., water [70, 128, 139]. The water content varies from 30 to nearly 100 % (w/v) [71]. Its close relation to biological tissue makes hydrogels extremely important for research and industrial applications [52]. The mechanical properties decide for which given application they are suitable, reaching from thickening agents in the food or cosmetic industry to tissue engineering or implants in the health sector. Its mechanical properties are usually mathematically described with theories for rubber or polymers [6]. The basis of a hydrogel are monomers which are linked to long chains. These polymers are physically or chemically cross-linked. The chemical linking is based on radical or condensation polymerization and results in very stable irreversible polymers [120] while physically linked polymers are a result of intermolecular forces such as hydrogen bonding, van der Waals forces or Coulomb interactions and entanglements of the long polymer chains [66]. The polymers are either chemically synthesized, e.g., poly-2-hydroxyethylmethacrylate, polyacrylic acid (PAA), polyvinyl alcohol (PVA) and silicon [53], or of biological nature, e.g., carrageen, alginate, agarose and gellan [63, 66].

2.2.1 Properties of Gellan and other Biopolymers

The interest in microbiologically synthesized biopolymers is constantly increasing [52]. According to Bajaj et al. [8] and Kreyenschulte et al. [82] xanthan, gellan and bacterial alginates are produced on commercial scale. Gellan is a linear, anionic polymer produced by different bacteria, e.g., *Pseudomas elodea* later reclassified to *Spingomonas elodea* as an extracellular polysaccharide (EPS). Gellan has a distinct gelation temperature and produces transparent gels [85, 156, 158]. In 1988 it was first approved for food usage in Japan but had already been used in cultivation mediums in biotechnology [118].

Fig. 2.5 shows the monomer structure of gellan gum which consists of a tetrasaccharide unit of →3)-D-glucopyranosyl-(β-1→4)-D-glucurono-pyranosyl-(β-1→4)-D-glucopyranosyl-(β-1→4)-L-rhamnopyranosyl-(α-1→ [114]. Ogawa and Kubota [123, 124, 158] determined the average molecular weight between $4.8 \cdot 10^4$ to $2.4 \cdot 10^5$ gMol^{-1}, while Bajaj et al. [8] reported a molecular weight of $M_n = 5.0 \cdot 10^5$ gMol^{-1}.

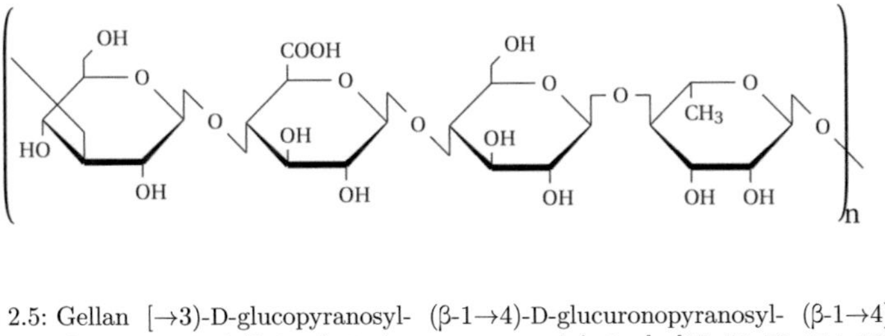

Figure 2.5: Gellan [→3)-D-glucopyranosyl- (β-1→4)-D-glucuronopyranosyl- (β-1→4)-D-gluco-pyranosyl-(β-1→4)- L-rhamnopyranosyl-(α-1→]$_n$ [35, 52, 70, 114, 128].

The persistence length was observed to be 98 nm at 25.0 °C and 17 nm at 40 °C [126, 158]. The swelling properties of gellan gum are influenced by the Na$^+$ ion concentration (c_{Na^+}). Low levels of Na$^+$ reduce the uptake of water. The functionality of gellan gum depends on the degree of acylation. Acylated gellan forms soft, very elastic, transparent and flexible gels while deacylated gellan results in hard, nonelastic brittle gels. Full hydration of deacylated gellan takes place between 80 and 95 °C, for acylated gellan 70 °C or higher is sufficient [114]. Gellation occurs for deacylated gellan between 10 and 60 °C depending on the concentration and presence of ions, while acylated gellan solidifies between 70 and 80 °C. Both gels show no hysteresis effect. Gelling is improved through the addition of acids and mono- and divalent ions for deacylated gellan, whereas the acylated gellan is not affected by these ions [114, 141]. Sugars in general reduce firmness and modify the texture of deacylated gellan gels, but increase overall strength of acylated gellan gels [114]. Fig. 2.6 shows the monomer structure of alginate consisting of β-D-mannuronic acid (M-block) and α-L-guluronic acid (G-block) [97, 136, 138].

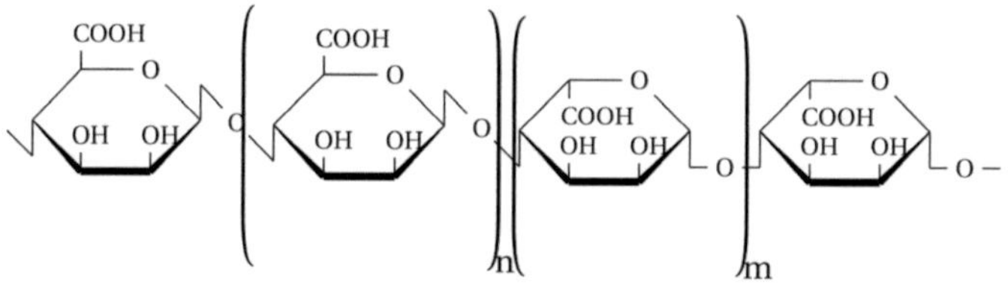

Figure 2.6: Alginate [β-D-mannuronic acid]$_n$ [α-L-guluronic acid]$_m$ [97, 136, 138].

Alginate forms in contrast to gellan heat-stable gels when reacting with Ca^{2+}. It has to contain a certain proportion of guluronic acid to form a gel. Junction zones between the alginate gel network are formed when a G-block of one alginate polymer is physically linked to a G-block in another alginate polymer. M-blocks and the MG-blocks do not participate in the junction zones but form so-called elastic segments in the gel network commonly visualized through the egg-box model as described by Grant et al. [56]. The egg-box model can be explained by energy considerations when looking at Lennard-Jones potentials, van der Waals forces and electrostatic interactions [15].

2.2.2 Gellan Gelling Mechanisms

The gelation process is generally explained as a two-step procedure. First it changes from a random coil to a double helix and then to helix-to-helix aggregate [51, 74, 156, 183, 184]. High acylated gellan forms double helix geometry at higher temperature than the deacylated gellan, causing weaker but more elastic gels with no hysteresis between melting and formation [122]. Thus, the acetyl groups prevent aggregation. According to Grant et al. [56] divalent metal ions promote the aggregation by site-binding between pairs of carboxyl groups on neighboring helices. This is in analogy to the egg-box model proposed for calcium-induced gelation of alginate and pectin [39]. Chandrasekaran et al. [23] showed that high quality deacylated polycrystalline gellan has a co-axial double helix with 3-fold symmetry formation in the solid state and confirmed that the egg-box model is sterically feasible [24]. Milas et al. [99] strengthened the proposed model of Chandrasekaran et al. [24] and described an aggregate with a helical confirmation.

Takahashi [158] stated in 1999 that the gelling mechanism was not fully understood. However, Valli et al. [66] supported in 2011 the gelling mechanism as proposed by Chandrasekaran and Thailambal in the late 1980's and early 1990's [114]. Structure formation of gellan gum is concluded in 24 to 36 hours. The viscoelastic properties of gels are strongly dependent on the connections between junction zones and the type of gel-promoting cation at the same gellan/cation ratio [106, 125, 174]. The gelation temperature and gel strength are influenced by ions and pH. As stated by Sworn et al. [156] Na^+ has a greater influence on the transition temperature than Ca^{2+} or Mg^{2+}, while the divalent ions increase the strength of the gel up to an optimum. This occurs at a concentration close the stoichiometric requirements of the carboxyl groups of the polymer (3 mM for 0.4 %(w/v) gellan). Monovalent metal ions bind to the surface of individual helices and lower their charge-density thus reducing the electrostatic barrier to aggregation [107]. The critical gelation concentration of K^+ is 33 % higher than for Na^+ at 0.5 % gellan, thus potassium ions are more effective in enhancing the gelling ability [68]. The critical gelation concentrations of Ca^{+2} and Mg^{2+} are almost equal but the increase in sol-gel temperature for the calcium ion is higher thus enhancing the gelling ability over magnesium [68]. Conclusively, viscoelastic properties of a gellan-based hydrogel can be adjusted by changing the ion concentration. Table 2.1 summarizes the published data, varying by several orders of magnitude for similar or even identical systems, e.g., Miyoshi et al. [100, 102] and Oliveira et al. [127].

Table 2.1: Values for storage (G') and loss (G'') moduli of gellan-based hydrogels with different ion concentrations (20 to 30 °C and a frequency of 1 s^{-1}). Data was obtained from figures and tables in referenced literature.

Gellan [% (w/v)]	Ions	G' [Pa]	G'' [Pa]	References
0.02	-	$5.4 \cdot 10^{-3}$	$1.2 \cdot 10^{-2}$	[129]
0.02	0.4 mM Ca^{2+}	$5.1 \cdot 10^{-2}$	$8.4 \cdot 10^{-3}$	[129]
0.02	1.2 mM Ca^{2+}	$8.9 \cdot 10^{-2}$	$4.1 \cdot 10^{-2}$	[129]
0.02	2.0 mM Ca^{2+}	$6.1 \cdot 10^{0}$	$2.1 \cdot 10^{-1}$	[129]
0.1	220 mM Na^{+}	$1.1 \cdot 10^{1}$	$1.0 \cdot 10^{0}$	[51]
0.125	102 mM Na^{+}	$6.6 \cdot 10^{-1}$	$2.9 \cdot 10^{-1}$	[156]
0.3	30 mM Na^{+}	$1.7 \cdot 10^{2}$	$1.2 \cdot 10^{1}$	[121]
0.3	100 mM Na^{+}	$1.5 \cdot 10^{3}$	$4.1 \cdot 10^{1}$	[121]
0.7	270 mM Ca^{2+}	$3.9 \cdot 10^{4}$	$3.2 \cdot 10^{3}$	[127]
1.0	-	$8.2 \cdot 10^{-2}$	$1.0 \cdot 10^{0}$	[102]
1.0	-	$3.7 \cdot 10^{-2}$	$7.9 \cdot 10^{-1}$	[161]
1.0	-	$7.5 \cdot 10^{-1}$	$1.5 \cdot 10^{-1}$	[100]
1.0	5 mM K^{+}	$3.2 \cdot 10^{-1}$	$1.0 \cdot 10^{0}$	[102]
1.0	10 mM K^{+}	$5.2 \cdot 10^{0}$	$6.8 \cdot 10^{0}$	[102]
1.0	30 mM K^{+}	$1.7 \cdot 10^{1}$	$8.7 \cdot 10^{0}$	[102]
1.0	50 mM Na^{+}	$1.5 \cdot 10^{-3}$	$3.7 \cdot 10^{-1}$	[161]
1.0	-	$2.0 \cdot 10^{2}$	$2.0 \cdot 10^{1}$	[122]
1.0	100 mM K^{+}	$1.0 \cdot 10^{3}$	$3.5 \cdot 10^{1}$	[122]
1.6	-	$8.1 \cdot 10^{-1}$	$7.0 \cdot 10^{-1}$	[117]
1.6	-	$3.5 \cdot 10^{2}$	$1.2 \cdot 10^{1}$	[92]
2.0	-	$8.6 \cdot 10^{0}$	$2.1 \cdot 10^{1}$	[100]
2.0	-	$1.9 \cdot 10^{1}$	$2.5 \cdot 10^{1}$	[102]
2.5	-	$6.9 \cdot 10^{1}$	$3.5 \cdot 10^{1}$	[100]
3.0	-	$5.9 \cdot 10^{1}$	$3.2 \cdot 10^{1}$	[102]
3.5	-	$1.3 \cdot 10^{3}$	$3.2 \cdot 10^{2}$	[100]

2.2.3 The Sol-Gel Temperature of Gellan-Based Hydrogels

Izumi et al. [68] published in 1996 phase diagrams and the molecular structure of gellan gum depending on the Na^+ concentration (c_{Na+}). They observed a difference between thermal transition and mechanical transition. The mechanical transition corresponds to the gel-sol transition while the thermal transition is a result of the melting of a local ordered structure. Additional studies describing the influence of Na^+ were published by Miyoshi et al. [101, 102]. They discovered true gel behavior below 30 °C and observed a two step transition. The sol-gel transition temperature increased with increasing salt concentration. However, Izumi et al. [68] described the thermal transition curve as insensitive to the type of salt but the mechanical transition to be very sensitive.

2.2.4 Oxygen Diffusion in Hydrogels

Oxygen solubility in water is poor but it is required as substrate and thus can limit growth. It is transported to the boundary layer of the hydrogel by the surrounding water or culture medium. A fast flow rate and water saturated with oxygen allow to assume that convective transport is not limiting. Thus, after passing the boundary layer the oxygen concentration within a hydrogel depends only on diffusion [142]. The oxygen transport to the immobilized surface (substratum) can be described by Fick's first law [49]

$$j = -D_F \frac{dc}{dx} \tag{2.30}$$

where j represents the flux, D_F the diffusion coefficient and dc/dx the concentration gradient in the direction of the surface. However, the time dependent measurement of oxygen profiles in the hydrogel cannot be described with Fick's first law, since it only considers a local dependency. To account for the dynamic behavior of oxygen diffusion into the hydrogel Fick's second law is required [49]

$$\frac{dc}{dt} = -D_F \frac{d^2c}{dx^2}. \tag{2.31}$$

Assuming the diffusion of the oxygen into an infinite hydrogel Eq. (2.31) can be solved to

$$c(x,t) = c^* - (c^* - c_0)\mathrm{err}\left(\frac{x}{2\sqrt{D_F t}}\right). \tag{2.32}$$

where c^* is the equilibrium concentration at the hydrogel surface which remains constant since the water is saturated with oxygen, c_0 is the concentration of oxygen at a depth x and the time t = 0. Oxygen diffusion in hydrogel was investigated by Hyust et al. [64] in calcium alginate, $\varkappa$-carrageenan, gellan gum, agar and agarose. They reported a reduction in oxygen diffusion with an increased polymer concentration and reported D_F values between $1.5 \cdot 10^{-9}$ and $2.1 \cdot 10^{-9}$ m^{2}s^{-1} for polymer contents of 0.5 to 5 %(w/v).

2.3 Continuum Mechanics

The field of continuum mechanics defines matters, the dimensions of which are several magnitudes larger than the molecular and atomic structure, as a continuum. Traditionally they are divided into two groups, solids and liquids. While solids have a defined volume and form, liquids have a defined volume but no defined form. The simplest description of the relationship between force and deformation of a linear elastic solid is given by

$$\sigma = G\gamma, \tag{2.33}$$

and called Hooke's law [17, 67], where σ is the stress, G the tensile or elastic modulus and γ the applied strain or shear. Fig. 2.7 shows the Hookean material, symbolized by a spring with G as spring constant.

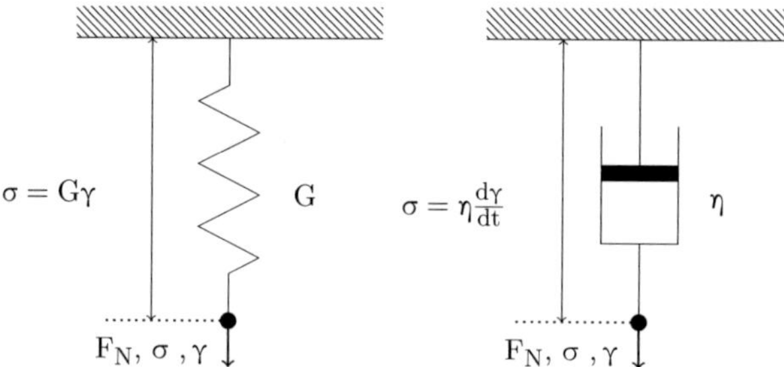

Figure 2.7: Hookean material (left) represented as spring and Newtonian material (right) represented as dashpot [167].

An ideal fluid behavior on stress can be described with Newton's law [162] as Eq. (2.34)

$$\sigma = \eta \frac{d\gamma}{dt} = \eta \dot{\gamma},$$ (2.34)

where η is the dynamic viscosity and $\dot{\gamma}$ the shear rate. This is often symbolized by a viscous damper (compare Fig. 2.7). However, most materials are neither ideal elastic nor ideal viscous but something in between, named viscoelastic [86].

2.3.1 Viscoelastic Model for Polymers

The relationship between stress and strain of viscoelastic materials depends on time. The materials differ in their response when exhibited to shear or extended and react not only with an instantaneous elastic deformation but also with a continuous viscous flow. The stress relaxation occurs in an exponential fashion. This is a typical linear elastic behavior for small strains ($\gamma \leq 0.5$) for polymeric liquids and is described with a relaxation modulus $G(t)$ [94]

$$G(t) = \frac{\sigma(t)}{\gamma}$$ (2.35)

with σ being the stress and γ the strain. After a certain time, the relaxation modulus reaches a plateau modulus G_e. Boltzmann suggested a general linear viscoelastic model [94] or a Prony series to describe $G(t)$ mathematically

$$G(t) = \sum_{i=1}^{n} G_i e^{-t/\lambda_i}.$$ (2.36)

According to te Nijenhuis et al. [162] three to five elements are usually sufficient for a single viscoelastic experiment to fully describe $G(t)$. Cross-linked rubbers show a short time relaxation followed by a constant modulus. Polymeric liquids show a behavior similar to rubber at shorter times with a nearly constant modulus plateau G_0, eventually followed by a flow at longer times [86, 160]. In oscillation experiments samples are loaded with a sinusoidal strain (γ), resulting in a by a phase angle (δ) shifted stress (σ) response or vice versa. The applied sinusoidal shear strain and the resulting stress response are given in Eqs. (2.37) and (2.38), respectively [86, 94, 162]:

$$\gamma(t) = \gamma_0(t)\sin(\omega t)$$ (2.37)

$$\sigma(\omega, t) = \gamma_0 \left[G'\sin(\omega t) + G''\cos(\omega t) \right],$$ (2.38)

where ω is the circular frequency and δ is the resulting phase angle. The storage (G'), loss (G''), and complex moduli (G^*) as well as the phase angle (δ) are defined in Eqs. (2.39) to (2.42)[162]

$$G' \equiv \frac{\sigma_0}{\gamma_0}\cos(\delta), \tag{2.39}$$

$$G'' \equiv \frac{\sigma_0}{\gamma_0}\sin(\delta), \tag{2.40}$$

$$G^* = G' + iG'' \tag{2.41}$$

and

$$\delta = \arctan\left(\frac{G''}{G'}\right). \tag{2.42}$$

Mechanical models often consist of spring and dashpot elements which resemble the typical Hookean or Newtonain behavior as discussed in Chapter 2.3 by Eq. (2.33) and Eq. (2.34), respectively. The most basic arrangements are the Maxwell- and the Voigt-Kelvin-model [162]. The first one consists of a dashpot and a spring arranged in series, while the latter uses both linked in parallel. Fig. 2.8 shows a model with n-Maxwell elements linked in parallel.

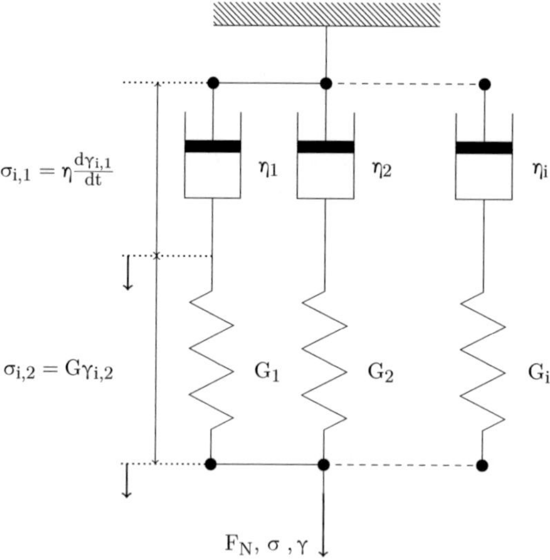

Figure 2.8: A Maxwell-Wiechert-model with n Maxwell elements in parallel [162].

This is also known as Maxwell-Wiechert- or the generalized Maxwell-model. The stress and strain relation for a Maxwell element can be written as

$$\sigma = \sigma_1 = \sigma_2 = \eta \frac{d\gamma_1}{dt} = G\gamma \tag{2.43}$$

and

$$\gamma = \gamma_1 + \gamma_2 \ . \tag{2.44}$$

The derivative of Eq. (2.44) with respect to time is

$$\frac{\gamma}{dt} = \frac{\gamma_1}{dt} + \frac{\gamma_2}{dt} = 0 \ . \tag{2.45}$$

The combination of Eq. (2.43) and Eq. (2.45) results in

$$\sigma(t) = \sigma_0 e^{-t/\frac{\eta}{G}} \ . \tag{2.46}$$

If compared to Eq. (2.36), Eq. (2.35) and with $\tau = \eta/G$ this can be rewritten to

$$G(t) = G_0 e^{-t/\tau} \ , \tag{2.47}$$

where τ is the relaxation time constant and $G(t)$ the relaxation modulus. For the n Maxwell elements in Fig. 2.8 linked in parallel Eq. (2.47) changes in analogy to Eq. (2.36) to

$$G(t) = \sum_{i=1}^{n} G_i e^{-t/\tau_i}. \tag{2.48}$$

This is equivalent to the result of the Boltzmann superposition principle and the Prony series [120, 140, 162] for polymers or the worm-like-chain-model [21, 41, 158] for polymers and biofilms. According to Macosko [94], te Nijenhuis et al. [162] and Ehret et al. [41] G' and G'' can also be written as functions of ω for n Maxwell elements

$$G'(\omega) = \sum_{i=1}^{n} G_i \frac{\omega^2 \tau_i^2}{1 + \omega^2 \tau_i^2} \tag{2.49}$$

and

$$G''(\omega) = \sum_{i=1}^{n} G_i \frac{\omega \tau_i}{1 + \omega^2 \tau_i^2} \ . \tag{2.50}$$

2.3.2 Worm-Like-Chain-Model

A gel can be seen as a highly hydrated solid with linear flexible polymer chains in an infinite solvent, hence water [6, 182]. The chains are connected by temporary bonds and have a characteristic length. For the worm-like-chain-model it is assumed that the chains have a defined end. Different types of attractions are considered to be electrostatic, dispersion forces, hydrogen bonds or entanglements giving the polymer stability [55, 78, 147]. Ehret et al. [41] assumed that all four mechanisms were presented in the EPS-matrix of biofilms resulting in four different bond types. These bonds are not considered permanent, thus are fluctuating, allowing for relaxation and creep behavior. This finally results in Eq. (2.49) and Eq. (2.50) with n = 4, where each Maxwell element represents one type of bond

$$G'(\omega) = \sum_{i=1}^{4} G_i \frac{\omega^2 \tau_i^2}{1 + \omega^2 \tau_i^2} \tag{2.51}$$

and

$$G''(\omega) = \sum_{i=1}^{4} G_i \frac{\omega \tau_i}{1 + \omega^2 \tau_i^2}. \tag{2.52}$$

2.3.3 Measurement Techniques

Liu et al. [90] described a polymer gel as a widely accepted material for tissue engineering, drug delivery and orthopedic load bearing. Its mechanical and physical properties can be easily adjusted to a wide range of applications. However, for soft gels accurate mechanical characterization is still challenging due to difficult sample preparation, fixture, gripping and load measurements. The research of Liu et al. [90] focused on nanoindentation techniques. The authors identified material parameters of polymer gels by inverse finite element techniques and used a nanoindentation creep test with a flat punch in combination with finite element simulations. Their results were correlated to an analytical solution based on a viscoelastic Kelvin-Voigt half-space model. To overcome limitations they used a 750 µm to 2 mm sized flat punch with a constant area [29]. However, the initial surface contact and the contact interface between surface and punch were problematic as well as the capillary attraction.

Furthermore, Liu et al. [92] used a rheometer to measure the storage and loss moduli of chitosan fiber enhanced gellan gum hydrogels. One major advantage of the rheometer as testing device over nanoindentation and other measurement techniques, such as tension [164] or compression devices [35, 84, 143] and magnetic microneedles [30] was the well defined deformation [12].

In classical viscoelastic materials testing creep, relaxation or oscillatory experiments are performed with a rotational rheometer [46]. In a plate-plate arrangement a sample is subjected to an oscillatory shear stress or strain. The displacement and stress are monitored and correlated to the applied time. Experimental results are then fitted to different rheological models, e.g., a (non linear) generalized Maxwell- [33, 36, 40], Burger- [36, 166, 167] or the worm-like-chain-model [21, 41].

Some of the first authors who published viscoelastic data on gellan gum were Sworn et al. [156] and Miyoshi et al. [100–102]. They used a standard rheometer (VOR, Bohlin Rheology, Lund, Sweden) and a dynamic stress rheometer (DSR, Rheometrics, N.J., USA) to analyze the structure development and gel-sol transition of gellan hydrogels during oscillatory shear stress. The ionic environment and gellan gum concentration had a big influence on the transition temperature, viscosity and structure development in the gellan gel system. Nishinari et al. [117] used a dynamic stress rheometer (DSR, Rheometrics, N.J., USA) to study the interaction between gellan gum and konjac glucomannan. Morris et al. [107] investigated the steady-shear viscosity (Sangamo Viscoelastic Analyser, Springfiled, IL., USA) using a cone plate configuration and Shimazaki et al. [141] analyzed as one of the first the influence of Ca^{2+} and Na^+ ions on the viscoelasticity of gellan gum in aqueous solutions. Comprehensive reviews on hydrogels and gellan were published by te Nijenhuis [161] and Nishinari [119]. While te Nijenhuis analyzed the thermoreversible hydrogels and their properties in general, Nishinari focused on the application and chemistry of gellan gum, in particular.

Current approaches to the viscoelastic characterization of hydrogels were performed by Nickerson et al. [113] who studied the viscoelastic properties of ϰ-carrageenan and alginate polysaccharides and the influence of Ca^{2+} and Na^+ ions similar to Oliveira et al. [127] and Nitta et al. [121]. Noda et al. [122] used an atomic force microscope (AFM) and related the results to the rheological behavior. They found a good agreement of rheological data with the structural information obtained using the AFM. The newest literature published in recent years by different authors describes the use of gellan gum or other hydrogels as biomaterial for tissue engineering [35, 92, 147]. They also used rheometric devices to characterize the viscoelastic properties.

2.4 Response Surface Methodology

The response surface methodology (RSM) provides a smooth visualization of information for problem optimization where the mechanisms that produce the data are either unknown or poorly understood. Thus, the mathematical form of the true response is unknown. An empirical linear or polynomial model (Eq. (2.53)) is fitted to the data and allows to predict the response within the design space [89, 144].

$$R = a_0 + a_1 x_1 + a_2 x_2 + ... + a_n x_n$$
$$+ a_{n+1} x_1 x_2 + ... + a_k x_n^m \tag{2.53}$$

where R is the response, $a_1...a_k$ are the polynomial coefficients and $x_1...x_n$ are the factors. The RSM of choice is the central composite design (CCD) which has been developed by Box and Wilson for chemical processes [54]. A CCD is set up with a finite number of factors, e.g., the concentrations of gellan, Mg^{2+} and Na^+ of the hydrogel. For visualization, the factors are limited to three factors or three dimensions, respectively. Then a CCD is presented as a cube, shown in Fig. 2.9 by the points #1 to #8, a center point (CP), and a star composed of the points #9 to #14. The cube, including the CP induces the first three, while the star adds another two levels of variation [144].

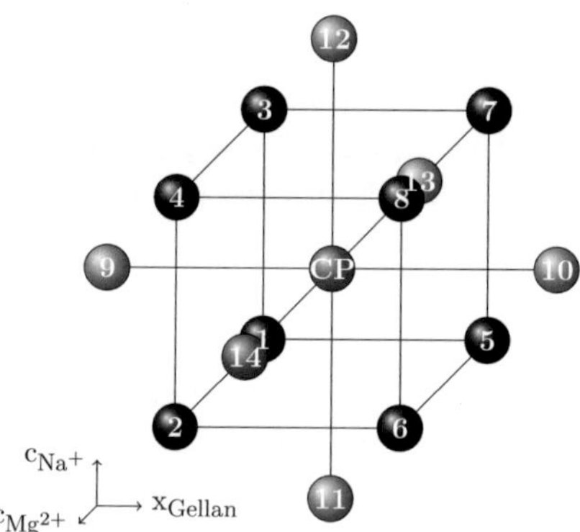

Figure 2.9: Scheme of the Central Composite Design (CCD) with three factors and three levels of variation represented by the cube (#1-#8, CP). This type of response surface methodology (RSM) introduces a fourth and fifth level of variation by introducing the star (#9-#14).

3 Material and Methods

3.1 Cultivation

Pseudomnas putida KT2440 is a Gram negative, rod-shaped (0.5 to 1.0 µm × 1.5 to 5.0 µm), saprotrophic soil bacterium which grows optimally at 25 to 30 °C [91, 176] and as biofilm [109, 116, 165]. It is able to metabolize e.g., aromatics like toluene, phenol or polystyrol. Thus, it is used in waste water treating or soil recovery processes [31, 43, 171]. Most *P. putida* strains are classified as pathogenic S2 microorganisms. However, the strain KT2440 is no carrier of virulence factors and is thus, considered to be safe and classified to the biosafty level S1 [112]. Growth experiments were either performed in 250 mL Erlenmeyer baffled shaking flasks, filled with 100 mL culture medium, at 30 °C and at a shaking rate of 120 min^{-1} with a lateral deviation of 50 mm in an incubator shaker (Certomat BS-1, Sartorius, Göttingen, Germany) as batch or in a biofilm tube reactor (BTR, see Chapter 3.1.5) also at 30 °C as continuous cultivation.

3.1.1 Cultivation Medium

The seed cultures were grown in the standard lysogeny broth (LB) containing 10 gL^{-1} Bacto-Trypton (Becton Dickinson, Franklin Lakes, NJ, USA), 5 gL^{-1} yeast extract (AppliChem, Darmstadt, Germany) and 10 gL^{-1} NaCl (Merck, Darmstadt, Germany). For the growth experiments in shaken flasks and the continuous BTR cultivations a low carbon source was used to promote biofilm formation but limit planktonic growth. The carbon source in the LB medium was reduced by a factor of 10 to 1 gL^{-1} Bacto-Trypton and 0.5 gL^{-1} yeast extract. All solid media were dissolved in deionized water, and the pH was adjusted to 7.

Additionally, an Autoinducer Bioassy 10 medium (AB10) adapted from Tolker-Nielsen et al. [165] was used. It was separated into stock solutions A and B, and a trace element solution. The final concentrations of solution A were 1.51 mM $(NH_4)_2SO_4$, 3.37 mM Na_2HPO_4, 2.20 mM KH_2PO_4, 179 mM NaCL, for solution B, 10 µM $CaCl_2$, 20.10 mM $MgCl_2$, 1 µM $FeCl_3$, 1 mM citrate, and for trace elements, 1.47 µM $CaSO_4$, 0.72 µM $FeSO_4 \cdot 7H_2O$, 0.12 µM $MnSO_4 \cdot H_2O$, 0.13 µM $CuSO_4$,

0.07 µM $ZnSO_4 \cdot 7H_2O$, 0.04 µM $CoSO_4 \cdot 7H_2O$, 0.05 $NaMoO_4 \cdot 7H_2O$, 0.08 µM H_3BO_3. Each solution was prepared with deionized water, the pH was adjusted to 7. All magnesium compounds and glucose were added filter sterilized to avoid precipitation after the broth was autoclaved at 121 °C for 20 min. For some experiments the amount of carbon in these growth media was altered, by changing the concentrations of either citrate or glucose.

The genetic modifications required the Super Optimal broth with Catabolite repression (SOC) [59]. It was used for the plasmid transformation and consisted of 20 gL^{-1} Bacto-Trypton (Becton Dickinson, Franklin Lakes, NJ, USA), 5 gL^{-1} yeast extract (AppliChem, Darmstadt, Germany), 10 mM NaCl (Merck, Darmstadt, Germany), 2.5 mM KCl, 10 mM $MgCl_2$, 10 mM $MgSO_4$ and 20 mM glucose in deionized water. The pH was set to 7, and everything autoclaved or filter sterilized.

If agar plates were prepared, 15 gL^{-1} agar was added to the medium required before being autoclaved. For selection purpose medium and agar plates were prepared with 1.25 mgL^{-1} tetracycline per 25 mL volume.

3.1.2 Seed Cultures

Seed cultures of *P. putida* KT2440 or *E. coli* K12 were cultivated overnight in Erlenmeyer baffled shaking flasks in LB medium as described in Chapter 3.1 and then used for characterizing growth experiments or concentrated for inoculation of the BTR. Concentration was done by centrifugation at 4000 min^{-1} for 10 min (Variofuge 3.0R, Heraeus, Hanau, Germany) and washing of the biomass and resuspension in a small amount of sterilized medium. The inoculum was taken from cryogenic cultures stored at -80 °C.

3.1.3 *P. putida* KT2440 and *E. coli* K12 Biofilms Grown on Membrane Filter*

The sample biofilms were based on *P. putida* KT2440 or *E coli* K12. In accordance with Körstgens et al. [80], biofilms were obtained by cultivating *P. putida* KT2440 or *E. coli* K12 overnight in LB medium at 30 °C in a shaking incubator. A bacterial culture was then diluted to a total of approximately 10^6 cells and filtered through a recycled cellulose membrane with an average pore size of 0.2 µm (Sartorius stedim

---

* Data was partillay obtained in cooperation with Vincent Wiegmann (2012) and Yvonne Kraus (2013) as part of their bachelor theses, TU Braunschweig.

Biotech, Göttingen, Germany). The filter membranes were placed on agar plates and cultured at 30 °C. After 48 hours, a mucoid bacterial biofilm lawn was obtained on the surface of the membrane filters, which were placed on the stainless steel plate of the rheometer (see Chapter 3.2.7) for viscoelastic characterization.

3.1.4 Hydrogel with Immobilized *P. putida* KT2440*

A seed culture was centrifuged at 4000 min^{-1} for 10 min (Variofuge 3.0R, Heraeus, Hanau, Germany) to concentrate the biomass. The supernatant was discarded and the pellet washed with Sörensen buffer [108] and again centrifuged at 4000 min^{-1} for 10 min, then resuspended in sterile distilled water to a volume of 3.75 mL with a final biomass concentration of approximately 5 % (w/v). 7.5 mL of a 1.5 %(w/v) gellan solution were mixed with the biomass suspension and carefully heated in a water bath to 47.5 °C. 120 µL of the 5 M NaCl solution and 40 µL 2.5 M MgSO$_4$ solution were added and carefully stirred for 10 min. As shown in Fig. 3.1 the mixture (B) was poured into the prepared mold consisting of coverglasses glued to the object slide (A). The cooled and solidified sample was stored at 4 °C until further usage. A sample of the hydrogel mixture was plated at the end of the treatment to ensure survival of the cells.

Figure 3.1: Object slide with in hydrogel immobilized cells: (A) Coverglasses glued to the object slide, (B) with hydrogel filled molding.

3.1.5 Cultivation in the Biofilm Tube Reactor

To guarantee reproducible and constant growth conditions a test site was assembled. Fig. 3.2 shows the P&ID flow diagram of the biological reactor test site. The system consisted of two biofilm tube reactors (BTR 1 and 2) first used by Horn [61] and changed by Möhle [103] and adopted for this work. A proposed modified BTR (not deployed) is discussed in Appendix A.1.

---

* Data was partillay obtained in cooperation with Vincent Wiegmann (2012) and Yvonne Kraus (2013) as part of their bachelor theses, TU Braunschweig.

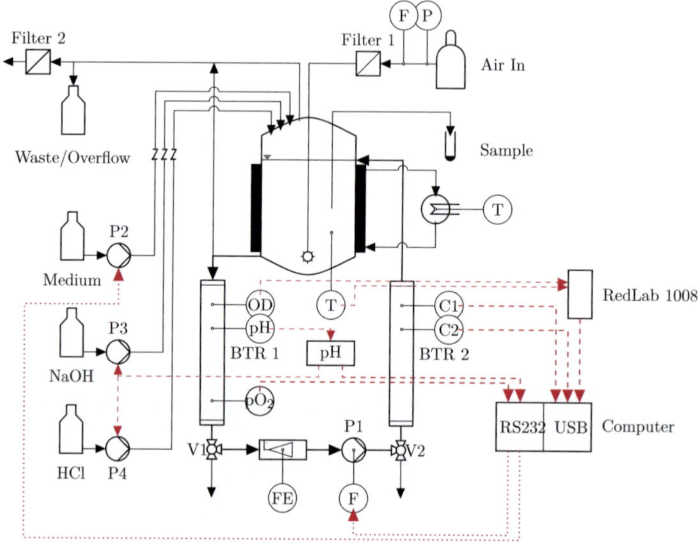

Figure 3.2: P&ID flow diagram of the test system with pump for the feed (P2) and for the pH control (P3 and P4), gassing chamber for oxygen saturation, Optical Density measurement (OD), pH-sensor (pH), temperature sensor (T), flow meter (FE) and oxygen sensor (pO$_2$). Two cameras (C1, C2) allow the observation of the biofilm development. Data acquisition is realized via an A/D converter (RedLab 1008), RS232 bus, USB and evaluation with Matlab. The pump (P1) is controlled with the software supplied by the manufacturer.

The BTRs have a diameter of 26 mm and were connected in serial by a pump (P1) (MDE 18 or MGD 15, Iwaki, Tokyo, Japan) controlled by a frequency converter (Combimaster CM25, Siemens, München, Germany or Varicon, Hanning Elektro-Werke, Oerlinghausen, Germany), to induce a constant circulating flow through the BTRs with 1.5 Lmin^{-1} and a Reynolds Number of approximately ~ 1500. Oxygen was provided by sparkling filter sterilized air through the medium in the gassing chamber at a constant flow rate of 1.2 Lmin^{-1}. The partial oxygen concentration was measured and kept close to saturation of 7.53 mgL^{-1} at 30 °C. The leaving air passed through a water trap. pH was kept constant via a pH controller (pH) (see Chapter 3.2.3). Data was collected with a RedLab 1008 AD converter via USB or RS232 connector and evaluated with a Matlab script.

After an initial batch phase of 0.5 to 2 days a periodical feed of 5 to 25 mL fresh media every 12 to 60 min was started. The fresh cultivation medium was added in small intervals to achieve a constant Optical Density (OD) (see Chapter 3.2.1) in the cultivation system. This resulted in a dilution rate of approximately $D = 0.5$ d^{-1}. The total volume of the biological reactor test site was approximately 1.2 L and kept constant via overflow through the air outlet. To reach a good contrast during image acquisition highly diluted cultivation media (see Chapter 3.1.1) were used. Thus, the cultivation medium of choice only allowed limited growth to an Optical Density of approximately $OD = 0.5$ to 1.0. The overall system can be described as a CSTR with an initial batch phase, therefore the balance equations for biomass and substrate are Eqs. (2.1), (2.2), (2.24), (2.25) and (2.27) to (2.29) as discussed in Chapter 2.1.6.

The inoculation was performed with a syringe up to an OD of 0.1. Other inoculation types depended on overgrown membrane filters and on cells immobilized in a gellan based hydrogel. Duration of the experiments varied from 3 to 17 days. Temperature throughout the experiments was kept at 30 °C, DO levels were targeted to be 67 to 100 %. pH was measured with a pH-electrode (SteamLine SL, Schott, Mainz, Germany) and kept at $pH = 7$.

3.1.6 Genetic Strain Modification*

The Green Florescences Protein (GFP) coding plasmid pSSBm85 (Fig. 3.3) was extracted from *Bacillus megaterium* YYBm1 and inserted into *P. putida* KT2440. Important genes are the selection markers for tetracycline (Tet^R) and ampenicillin (Amp^R), which result in a resistance of *P. putida* KT2440 to these antibiotics. The transcription was regulated via a repressor (XylR) and the binding region (XylR-bind) which allowed for a xylose induced GFP production [145].

For the extraction process of pSSBm85 plasmid a NucleoBond Xtra Midi Kit (Macherey-Nagel, Düren, Germany) and the low-copy plasmid purification protocol of Nagel were used [93]. *B. megaterium* YYBm1 was cultured overnight at 30 °C at 120 min^{-1} in LB media with added fructose and tetracycline. After centrifugation at 4000 min^{-1} for 10 min (Variofuge 3.0R, Heraeus, Hanau, Germany) and washing, the pellet was resuspended in a special RNase A containing buffer which was part of the kit. The addition of a lysis buffer destroyed the cellular walls of *B. megaterium* YYBm1 and allowed the extraction by precipitation and filtration of the plasmid.

---

* Data was partially obtained in cooperation with Florian Geppert (2012) as part of a student research project, TU Braunschweig.

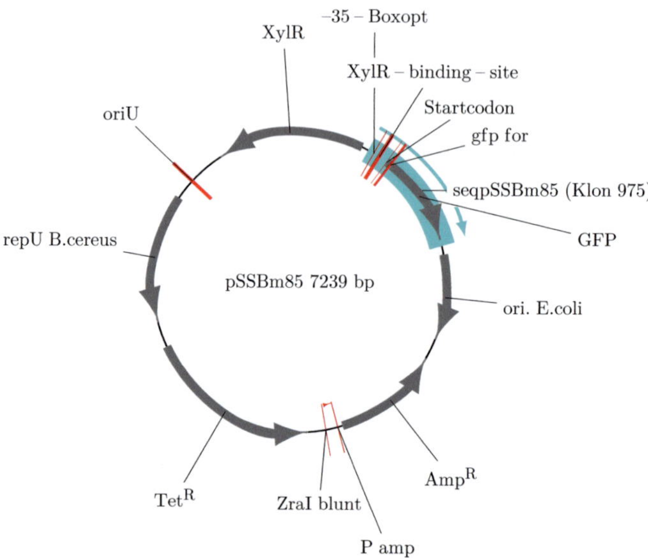

Figure 3.3: Plasmid extracted from *B. megaterium* YYBm1 inserted into *P. putida* KT2440. Important genes are marked, including selection markers, derived from *Bacillus cereus* for tetracyclin (Tet^R) and from *Escherichia coli* for ampenicillin (Amp^R), transcription regulation via the XylR repressor and the binding region (XylR-bind) to allow a xylose induced GFP production [145].

After elution and refiltration, reprecipitation and multiple washing and centrifugation steps at 8500 min^{-1} for 30 min, at 4 to 20 °C (Biofuge Stratos, Heraeus, Hanau, Germany) the plasmid pellet was dissolved in distilled water and stored at -20 °C. The gene transfer to *P. putida* KT2440 was reached via electroporation. *P. putida* KT2440 was cultured overnight at 30 °C at 120 min^{-1} in SOC medium. The culture broth was centrifuged and washed, the pellets were resuspended in 15 % glycerin and stored at -80 °C. An electroporation chamber (Bio-Rad Laboratories, Munich, Germany) was prepped with a special electroporation cuvette holding 30 µL of cell and 2 µL plasmid suspension. After the electro-magnetic pulse had been applied the cell solution was mixed with fresh SOC or LB medium and cultured at 30 °C for 60 min at 200 min^{-1} before being plated to tetracycline containing agar plates. After 48 hours at 30 °C small colonies of transformed *P. putida* KT2440 had been formed. These were than recultivated in LB medium and stored as cryogenic cultures for further usage.

3.1.7 Sample Preparation of Biomimetic Hydrogels

Gellan (Gelrite) was obtained from Carl Roth and was produced by Merck (Kelco Division, Rahway, NJ, USA). The gellan was already highly purified thus no further purification was required. Prior to sample formulation, stock solutions of gellan and the chloride and sulfate salts of Na^+ and Mg^{2+} had been prepared. The gellan stock solution contained 1.5 % (w/w) gellan and deionized water and was heated to 90 °C and maintained at this temperature for 60 min to ensure complete dissolution and hydration of the polysaccharides. The stock solution was stored as a liquid at 60 °C. The chloride and sulfate salts of Na^+ (Merck, Darmstadt, Germany) and Mg^{2+} (Sigma-Aldrich, St. Louis, USA) were separately dissolved in deionized water to prepare 5 M and 2.5 M stock solutions, respectively. To prepare the hydrogel samples, deionized water was preheated to 70 °C. Depending on the desired final concentrations of gellan, Na^+, and Mg^{2+}, appropriate amounts of the respective stock solutions were added to a beaker, and the preheated deionized water was added to achieve a final total volume of 30 mL. The solution was then stirred for 10 min at 70 °C. The beaker was covered to prevent evaporation. Fig. 3.4 shows the 2 mL volume of the liquid solution, which was poured into a cylindrical mold ($\varnothing$ = 40 mm, height = 2 mm) and placed on aluminum foil; this process was repeated with additional molds to obtain multiple samples. After 10 min, the gelled hydrogels were placed into Petri dishes, which were sealed to prevent evaporation and stored at 5 °C for 24 h. Fig. 3.4A shows the mold, Fig. 3.4B presents some hydrogel samples with and without the mold placed on aluminum foil. Fig. 3.4C depicts the hydrogel arranged on the rheometers bottom plate.

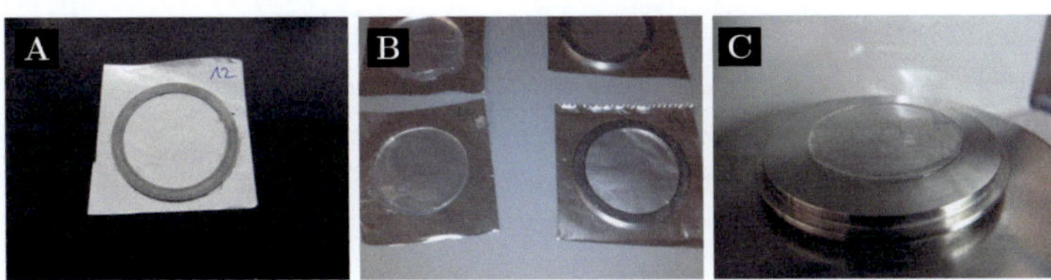

Figure 3.4: (A) Mold for the hydrogel, (B) hydrogel with and without mold, (C) hydrogel placed on the rheometer.

3.2 Analytics

The continuous cultivation in the BTR was steadily monitored and dissolved oxygen (DO), pH, temperature and Optical Density (OD) were logged with Matlab (Math-Works, Version 8.1.0.604) or the software provided from the hardware supplier.

3.2.1 Calibration of Optical Density Sensor

The used OD sensor was constructed at the Institute of Biochemical Engineering, TU Braunschweig (D. Rasch). It utilized a LED (600 nm) and a photo transistor which were placed on opposing sides on the glass tube. The passing through light was correlated to a voltage via the transistor and digitalized by a DAQ device (ME-RedLab 1008, Meilhaus Electronic, Munich, Germany). Matlab was used to record the data from the A/D converter. The voltage (U), the OD in the reactor, the intensity of the LED (I) and the OD measured with a photometer were correlated during calibration, resulting in Eq. (3.1) ($R^2 = 0.991$),

$$OD = 2.008 \cdot e^{(0.727 \cdot I - 6.474) \cdot U} . \tag{3.1}$$

However, environmental diffusive light disturbed the quality of measurements. Thus the sensors and the tube reactor were generously covered with aluminum foil.

3.2.2 Correlation of Optical Density and Bio Dry Weight

The Optical Density (OD) was correlated to the bio dry weight (BDW) by taking samples of cultivations. The OD was estimated with a photometer against the used media, then the 5 mL samples were washed three times and resuspended. The samples were piped into dried and weighted test tubes which were centrifuged again. The supernatant was discarded and the test tubes with the pellets were dried at 60 °C for 48 hours. The difference in mass was equivalent to the weight of the biomass in 5 mL culture broth. The correlation is given in Eq. (3.2) ($R^2 = 0.81$)

$$BDW = 0.658 \cdot OD . \tag{3.2}$$

3.2.3 Calibration of pH Sensor

The pH sensor (SteamLine SL, Schott, Mainz, Germany) and controller (pH Control 2, Meredos, Bovenden, Germany) included a PD device. The settings employed are given in Table 3.1. 0.1 M NaCl and HCl solutions were used to keep the pH constant and added via two peristaltic pumps (P3 and P4, SP-GV, Meredos, Bovenden, Germany, compare Fig. 3.2 with a flow rate of 1.6 mLmin^{-1}.)

Table 3.1: Parameters used for the pH controller

Parameter	Value	Function
pH_{min}	6.85 - 6.95	minimal set point
pH_{max}	7.05 - 7.15	maximal set point
pH_{hy}	0.25 - 1.00	overshoot to prevent oscillation
dP	0.01 - 13.99	interval of proportional control
dt	10 - 100 s	duration between pulses
Pt	5 - 10 s	duration of pulse

3.2.4 Calibration of Temperature Sensor

The temperature was kept constant at 30 °C in all continuous cultivations. Therefore, a jacked reactor was used. A temperature sensor (GPRT 1400 AN, Greisinger electronic, Regenstauf, Germany) was placed inside the reactor and the voltage signal recorded with Matlab and correlated to the temperature. A thermostat (K12 KS, Lauda, Lauda-Königshofen, Germany) was connected to the reactor jacket and controlled manually. A heating temperature of 33 °C was required to hold the temperature inside the reactor at the set point of 30 °C. The calibration for the temperature is given in Eq. (3.3)($R^2 = 0.997$)

$$T = 104.7 \cdot U \, , \tag{3.3}$$

which is close to the 10 mV per °C provided in the manual.

3.2.5 Calibration of Dissolved Oxygen Sensor

Oxygen supply was achieved by bubbling air through a glass frit into the medium. The air was filter sterilized at the in- and outlet. Dissolved oxygen levels were measured with pO_2 electrodes (Oxical 325,WTW, Weilheim, Germany or VisiFerm DO, Hamilton, Bonaduz, Switzerland) and monitored with the Oxi 597-S (WTW, Weilheim, Germany) or the software supplied by Hamilton. The Oxi 597-S was connected to the PC via ME-RedLab 1008 (Meilhaus Electronic, Munich, Germany). The saturation concentration of the air oxygen in water at atmosphere and 30 °C was set to 7.53 mgL^{-1} [42].

3.2.6 Oxygen Microelectrode Measurements

Fig. 3.5 shows a section of the BTR, the microelectrode was attached to the micromanipulator and inserted through the port to measure the oxygen profiles in the hydrogel. The pO_2 microsensor (Fig. 3.5, A,B)(ox-10, Unisense, Aarhus, Denmark) was connected to a picoammeter (PA2000, Unisense, Aarhus, Denmark) and the micromanipulator (Fig. 3.5, C) to a controller (Oriel Mike Encoder 18011, Newport Spectra-Physics, Darmstadt, Germany). The output of the picoammeter was recorded with a ME-RedLab 1008 DAQ device (Meilhaus Electronic, Munich, Germany) and evaluated with Matlab. The micromanipulator was also controlled with Matlab, this allowed a programmed measurement at certain hydrogel depths.

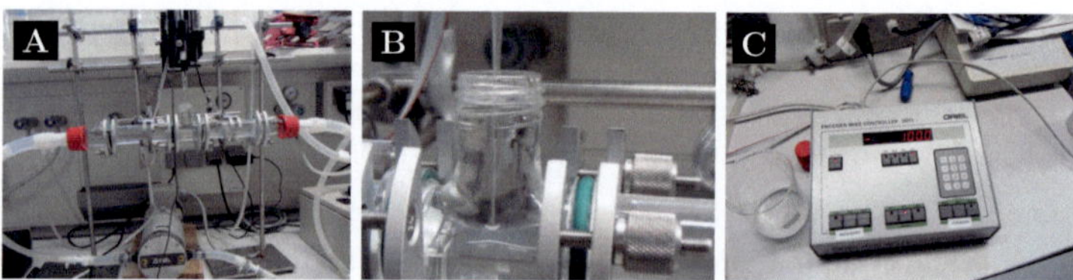

Figure 3.5: Dissolved oxygen measurement in hydrogels with microelectrodes. A: Section of the test site, B: Zoomed in on section of BTR with microneedle, C: micromanipulator controller.

3.2.7 Rheology

The rheological parameters storage (G$'$) and loss (G$''$) moduli were measured with a plate-plate Bohlin Gemini II rheometer (Malvern Instruments, Worcestershire, UK) at 25 °C in cooperation with the Institute for Particle Technology, TU Braunschweig (S. Günther). The frequency was set from 0.1 to 10 s^{-1}. At frequencies lower than 0.1 s^{-1}, the water stored in the hydrogel evaporated rapidly, resulting in false measurements. The tensile stress (σ) was maintained within a range of 0.5 to 500 Pa and an oscillatory strain of $\pm\, 5 \cdot 10^{-4}$ was applied, limited to the linear range of the hydrogels. Data was obtained at 25 logarithmic scaled frequencies with a resolution of 11 bits and averaged over 2 s. The normal force was kept constant within a range of 1 ± 0.3 N. Fig. 3.6 shows the hydrogel placed on the bottom plate (Fig. 3.6, A), between both plates (Fig. 3.6, B) and during the measurement (Fig. 3.6, C).

Figure 3.6: Hydrogel measurement with plate-plate. A: Hydrogel placed on bottom plate, B: Hydrogel between both plates, C: Hydrogel during measurement.

3.3 Response Surface Methodology

To study the influence of Na^+ and Mg^{2+} on the viscoelastic properties of different gellan-based hydrogels within a frequency range of 0.1 to 10 s^{-1}, a response surface methodology (RSM) was used. The response surface of choice was the central composite design (CCD). The concentration of gellan and the ions resulted in three independent factors, the frequency in an additional fourth factor and the viscoelastic parameters in terms of storage (G') and loss (G'') moduli provided two independent responses. The complex modulus (G^*), compare Eq. (2.41), the phase angle (δ), compare Eq. (2.42), and the complex viscosity (η^*) can be computed from G', G'' and ω, thus they are no independent responses and not considered further. Hence, the CCD was set up with the already mentioned four factors and five levels of variations. Eq. (3.4) shows the quadratic model for four factors: the content of gellan (x_{Gellan}), the concentration of Mg^{2+} ($c_{Mg^{2+}}$) and Na^+ (c_{Na^+}), as well as the logarithmic frequency (f)

$$R = a_0 + a_1 x_{\text{Gellan}} + a_2 c_{\text{Mg}^{2+}} + a_3 c_{\text{Na}^+} + a_4 f$$
$$+ a_5 x_{\text{Gellan}} c_{\text{Mg}^{2+}} + a_6 x_{\text{Gellan}} c_{\text{Na}^+} + a_7 x_{\text{Gellan}} f$$
$$+ a_8 c_{\text{Mg}^{2+}} c_{\text{Na}^+} + a_9 c_{\text{Mg}^{2+}} D + a_{10} c_{\text{Na}^+} f$$
$$+ a_{11} x_{\text{Gellan}}^2 + a_{12} c_{\text{Mg}^{2+}}^2 + a_{13} c_{\text{Na}^+}^2 + a_{14} f^2. \tag{3.4}$$

The softwares used are Design-Expert (Stat-Ease, Version 7.1.6), Matlab (Math-Works, Version 8.1.0.604) and OriginPro (OriginLab Corporation, Version 9.0.0 (64bit) SR2). Statistical values are provided by the used design of experiments software Design Expert. The quality of the fitted model is described with the F-value, R^2 and Adequate Precision Ratio.

3.3.1 Statistical Methods

In order to ensure the quality of the model and the data, statistical approaches were required to verify the results obtained. A Box-Cox analysis was used for power transformation and to improve the results by adjusting the model. The approach assumes, that if the computed model is adequate to fit the data, its residuals should be normally distributed since they are caused by random error. The residuals should have a mean close to zero. With help of a Box-Cox analysis not normally distributed data was transformed by the power of λ. λ is usually between -3 and 3 in steps of 0.5.

$$R' = \begin{cases} R^\lambda & \text{if } \lambda \in \mathbb{R} \backslash 0 \\ \log(R) & \text{if } \lambda = 0 \end{cases} \tag{3.5}$$

The Box-Cox analysis computes the minimal residual sum of squares as a function of λ and provides a guide-line for selecting the correct power law transformation. The optimal value of λ yields the minimal residual sum of squares. The transformed data can improve the quality of the model but might not always result in normally distributed data [144]. To ensure the normal distribution of data, a Gaussian distribution of standard residuals was used to visualize the density of the residuals (compare Fig. 4.27).

Some additional statistical values were provided by the Design of Experiments Software Design Expert. The quality of the fitted model is described by the F-value, R^2, and Adequate Precision Ratio. The F-value compares the variance of the model with the variance of the residuals; the larger the F-value, the higher the significance of the model. R^2 is a measure of the amount of variation around the mean. Finally, the adequate precision is a ratio of the signal to noise. It compares the predicted values of the design points to the average prediction error. Ratios greater than four are considered adequate for model discrimination [10, 54, 144].

4 Results and Discussion

To engineer a suitable and reliable biofilm mimetic hydrogel the characteristics of the model organism *P. putida* KT2440 and at least the biofilm properties had to be known. Therefore, the biological test site to cultivate and test either the biofilm or the hydrogel had to be established (compare Chapter 3.1.5) and a proper reactor design for controlled growth and desired analytics had to be constructed. The first was mostly automated to allow long time cultivations and constant measurements, the latter was based on the design of Horn [61] and Möhle [103], who developed a biofilm tube reactor (BTR). However, its design had some limitations, thus was further optimized for later use with surface recognition techniques such as CLSM or stereoscopic vision but eventually not implemented (compare Appendix A.1). Additionally, the hydrogel composition had to be adjusted to imitate the desired physico-chemical properties of the hydrogel. This was done with a response surface methodology, a design of experiment approach. Finally, the hydrogels were used in experiments as biofilm substitutes to proof their similar characteristics in comparison to real biofilm samples.

4.1 Growth of *Pseudomonas putida* KT2440 on Different Media

In the beginning, the growth characteristics of *P. putida* KT2440 were analyzed. Sessile growth often is a result of some kind of substrate limitation (compare Chapter 2.1), therefore the goal of these growth experiments was to limit nutrient availability without starving the bacteria to death. As suggested in literature carbon sources were either glucose or citrate [57, 69, 116]. The media compositions are provided in Chapter 3.1.1. The concentrations of trypton and yeast extract were reduced in steps of 10 (1/1C, 1/10C, 1/100C) resulting in 10 gL^{-1} to 1.0 gL^{-1} and finally 0.1 gL^{-1} tryptone and 5 gL^{-1} to 0.5 g^{-1} and 0.05 gL^{-1} yeast extract, respectively. Citrate concentrations were 10 mM, 4 mM and 1 mM as used by Tolker-Nielsen [165]. Additionally, an AB10 medium with 10 mM glucose was tested. Fig. 4.1 shows the growth of *P. putida* KT2440 on LB and AB10 medium in batch cultures at 30 °C.

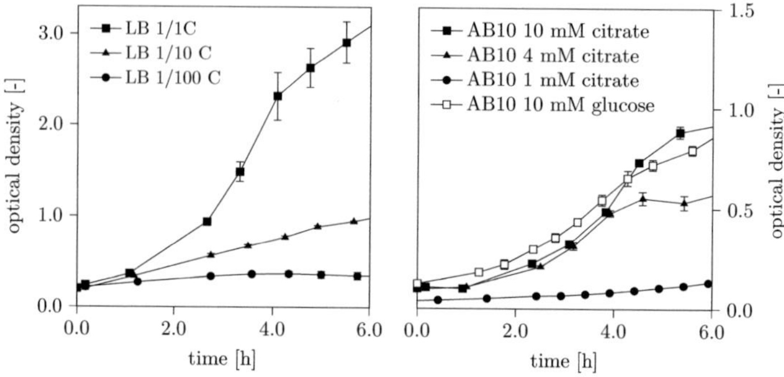

Figure 4.1: Growth of *P. putida* KT2440 on LB (left) and AB10 medium (right), at 30 °C, 120 min^{-1}, 100 mL volume in 250 mL shaken flasks, at different concentrated carbon sources and an initial OD of 0.1.

All diagrams showed exponential growth if substrate concentrations were sufficient. Growth on LB medium increased from a specific growth rate of $\mu = 0.15$ h^{-1} ($R^2 = 0.98$) and $\mu = 0.30$ h^{-1} ($R^2 = 0.97$) to $\mu = 0.59$ h^{-1} ($R^2 = 0.99$) with increasing substrate concentration. If cultured on LB at a concentration of 1/100C the OD hardly increased for 6 hours. This was also reflected in the low growth rate of $\mu = 0.15$ h^{-1}. The final OD reached on LB medium was close to 3.1, while on citrate growth was limited to an OD below 1.0. Growth on citrate at a concentration of 1 mM resulted in a decaying OD first, but increased after 6 hours. This, however, was most likely due to the autolysed cell fragments that were metabolized later resulting in a slight increase in growth. Thus, a concentration of 1 mM of citrate was not sufficient to support growth survival and maintain growth. For the medium with 4 mM and 10 mM citrate the specific growth rate μ increased from $\mu = 0.43$ h^{-1} ($R^2 = 0.98$) to $\mu = 0.47$ h^{-1} ($R^2 = 0.96$) with raising substrate concentration. Growth on AB10 medium with 10 mM glucose reached $\mu = 0.40$ h^{-1} ($R^2 = 0.99$) and was comparable to the growth on 10 mM citrate. The experiments on media with 4 or 10 mM citrate or 10 mM glucose and the LB medium provided all sufficient growth. The obtained specific growth rates were comparable to the data published by Jahn et al. [69] who determined $\mu_{max} = 0.34$ h^{-1} on 10 gl^{-1} citrate and of $\mu_{max} = 0.28$ h^{-1} on 10 gl^{-1} glucose in batch at 30 °C (compare Chapter 2.1.2). As anticipated, maximum growth was measured on standard LB medium, but a high OD hindered an observation of the biofilm development via image acquisition techniques. Thus, the LB medium was only used diluted.

4.2 GFP Producing Mutant of *P. putida* KT2440*

The successful insertion of the pSSBm85 plasmid into *P. putida* KT2440 was tested with a SDS-page. Fig. 4.2 shows the gene fragments of the host organism (*B. megaterium*), a wild type reference strain of *P. putida* KT2440 (Ref), the standard marker (Marker) and the *P. putida* KT2440 mutant grown in SOC (SOC) and LB medium (LB). The region of interest (ROI) was between 20 and 25 kDa and is marked accordingly.

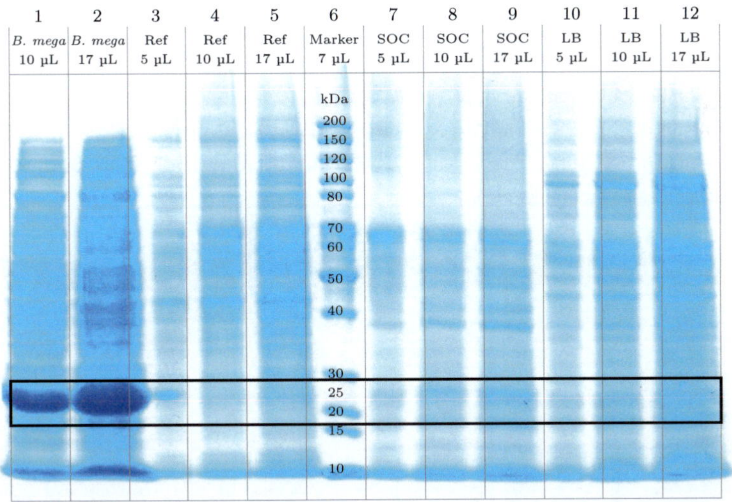

Figure 4.2: Result of transformation. The SDS-page shows the target gen fragments at 20 to 25 kDa compared to the host (*B. megaterium*) and the clones (*P. putida* KT2440) on LB and SOC medium.

Column 6 (Marker, 7 µL) shows the markers which are clearly separated, signaling a good general separation of the fragments. The SDS-page developed intense bands for the desired fragments of the host bacterium, which was a result of high amounts of plasmid in these cells. The reference strains of *P. putida* KT2440 showed only a slight blueish color, probably a consequence of other fragments or noise. The columns marked SOC and LB (7 to 9 and 10 to 12, respectively) are the results

---

* Data was partially obtained in cooperation with Florian Geppert (2012) as part of a student research project, TU Braunschweig

for the new mutant strain and color intensity in the ROI increased with sample concentration. The transformed *P. putida* KT2440 mutant strain seemed to favor the LB medium for the plasmid production. Even though the intensity of the color was notably weaker when compared to *B. megaterium* bands it still showed a working genetic transformation.

Fig. 4.3 presents images captured with a CLSM (Institute of Plant Biology, R. Hänsch, TU Braunschweig) at different resolutions and shows the working GFP expression (green) of a *P. putida* KT2440 culture. Additionally, Syto 9 (red) was used as stain for dead cells. It is clearly visible that the GFP producing cells were living and not identical with the red dead cells. Obviously, most cells in the sample produced GFP with only a small proportion of dead cells. In future work the new GFP mutant of *P. putida* KT2440 can be used to reconstruct biofilm volume models from the biofilms as shown by Böl et al. [13] and to visualize biofilm development over time. The obtained data facilitates to verify the physico-chemical and theoretical biofilm models.

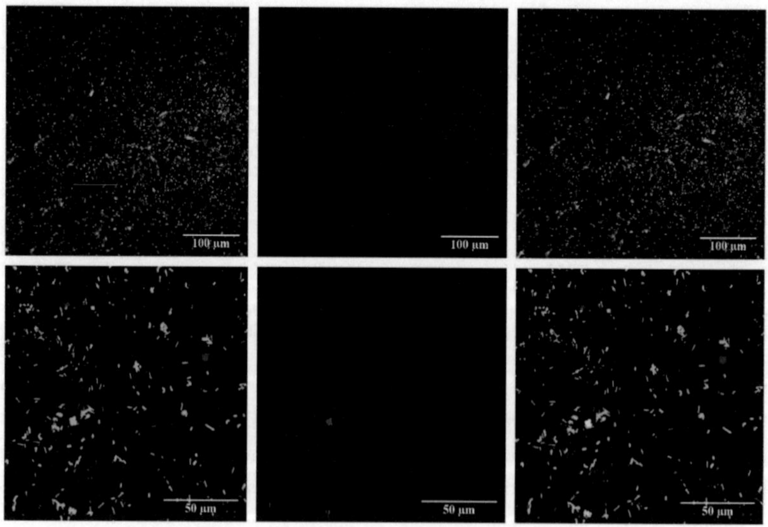

Figure 4.3: *P. putida* KT2440 with GFP (green, left), Syto 9 stained dead cells (red, center), merged images with GFP and Syto 9 (right).

4.2.1 Growth of *P. putida* KT2440 pSSBm85*

For the further usage of the mutant strain it was of interest to compare its growth characteristics with those wild type. Therefore, the mutant strain was cultured on the AB10 medium with 4 mM citrate and both with and without addition of xylose D(+) which induced GFP production. Fig. 4.4 shows the growth for both cultivations with an initial OD of 0.1. After 7 h the maximal OD reached approximately $OD_{max} = 0.3$ which is below $OD_{max} = 0.5$ for the wild type strain discussed in Chapter 4.1.

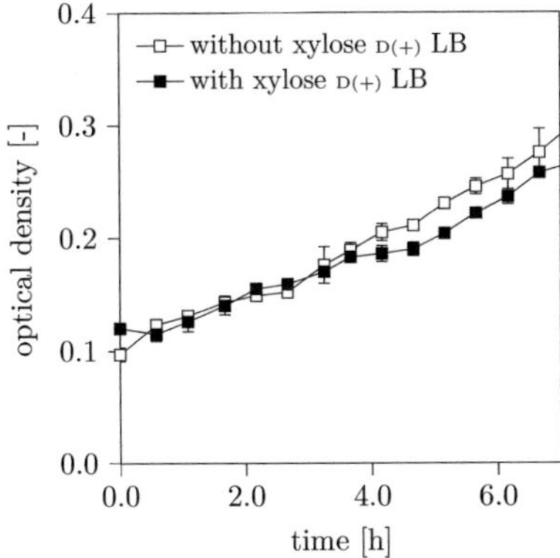

Figure 4.4: Growth of *P. putida* KT2440 pSSBm85 mutant on AB medium with 4 mM citrate as carbon source and with 4 gL^{-1} xylose D(+) as well as without xylose D(+). The specific growth rates are similar with $\mu = 0.14\,\mathrm{h}^{-1}$ ($R^2 = 0.993$) and $0.12\,\mathrm{h}^{-1}$ ($R^2 = 0.986$), respectively.

The specific growth rate only changed slightly from $\mu = 0.12\ \mathrm{h}^{-1}$ with xylose D(+) to $\mu = 0.14\ \mathrm{h}^{-1}$ without xylose D(+) which is insignificant, however, compared to the wild type dropped by 67 % (compare Chapter 4.1). Possible reasons were an energy intense production of the GFP or the plasmid reproduction which lowered the energy available for growth.

---

* Data was partially obtained in cooperation with Florian Geppert (2012) as part of a student research project.

4.3 BTR Cultivation and Biofilm Development

The first cultivation experiments were performed to verify the working test system and the data acquisition. As stated in Chapter 3.1.1 a standard AB10 medium according to Tolker-Nielsen et al. [165] with 4 mM citrate or 10 mM citrate was used. Later, this medium was modified and citrate replaced by 10 mM glucose to enhance growth. Additionally, cultivations on LB medium with 1.0 gL^{-1} tryptone and 0.5 gL^{-1} yeast extract were performed. Furthermore, the inoculation type was changed to force biofilm growth on specific surfaces. These were either the BTR elements, silicon or glass tube walls, wire and plastic nettings, a gellan based hydrogel with immobilized cells or filter membranes. Bacterial growth was monitored via online OD sensors, which had been calibrated ahead as shown in Chapter 3.2.1. Some culture samples were taken occasionally and the BDW was measured to verify data acquisition. Table 4.1 summarizes the different cultivation experiments.

Table 4.1: Cultivations in BTR

Feed carbon source	Replicates	Comments
4 mM citrate	3	seed culture
10 mM citrate	3	seed culture & overgrown filter
10 mM glucose	3	immobilized cells & overgrown filter
10 mM glucose	3 (1)	seed culture
yeast/trypton	3	overgrown filter

4.3.1 Cultivation in AB10 Medium with 4 mM Citrate

The experiments were performed on AB10 medium with 4 mM citrate as carbon source. Temperature was set constant to 30 °C, pH was kept at 7 and targeted dissolved oxygen level was 100 %. The PD pH controller settings were set to $pH_{min} = 6.85$, $pH_{max} = 7.15$, $pH_{hy} = 0.25$, dp = 1, dt = 100 s and Pt = 5 s. Inoculation was accomplished with a concentrated seed culture of *P. putida* KT2440 (compare Chapter 3.1.2), which was injected into the system with a syringe to an initial OD of 0.1. Fig. 4.5 shows the averaged results of the experiments done as triplicate. The first two days were characterized by a batch phase with a specific growth rate of $\mu = 1.18$ d^{-1} which was by factor 9 lower than in the shaken flask experiments in Chapter 4.1 on AB10 with 4 mM citrate with a growth rate of $\mu = 10.32$ d^{-1}

on the same cultivation medium. However, both experiments reached a maximal OD_{max} of 0.5. After the batch phase the feed was initiated with $F = 0.6$ Ld^{-1}. This resulted in a dilution rate of $D = 0.5$ d^{-1} which was smaller than the critical dilution rate of $D = \mu$ to avoid wash out, compare Eq. (2.27) in Chapter 2.1.6.

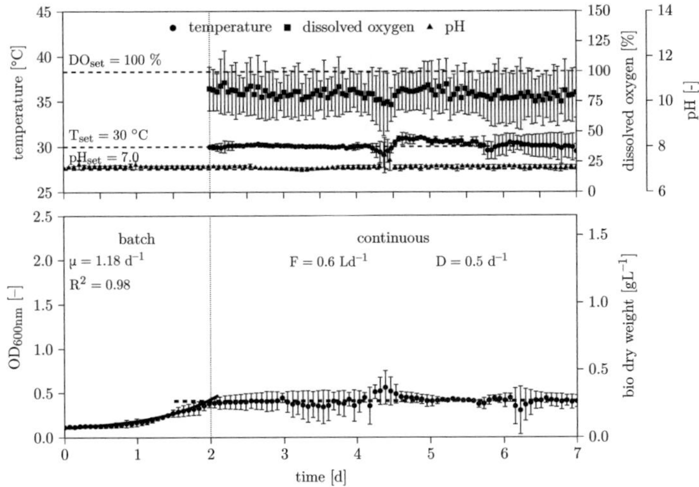

Figure 4.5: Cultivation experiment of *P. putida* KT2440 in the BTR on AB10 medium with 4 mM citrate as carbon source. The first 2 days were operated as batch, followed by a continuous phase ($D = 0.5$ d^{-1}). pH was set to 7, temperature to 30 °C and targeted oxygen levels were close to 100 % saturation.

Thus, the constant OD level was the result of the limited substrate availability. The temperature level during the experiments only deviated a little from the set point. However, during the fifth and seventh day of cultivation an increase in the error bars was recognizable. This was most likely a result of the change of the stock medium bottle for the feed. The targeted oxygen level of 100 % could not be reached, but higher gassing rates resulted in increased foaming. Thus, oxygen availability was reduced to an average 80 %, which in general does not limit growth. The overall error in the oxygen concentration was high, which was caused by a poor O_2 sensor (WTW, Oxical 325). The relatively high errors are typical for a biological system since environmental influences are in general eminent. However, the final objective was to obtain single species biofilms that were stressed identically and grown under constant conditions. The biofilm growth in these experiments was not satisfying but the automated measurements of OD, DO, pH and temperature showed acceptable results. The PD controlled pH regulation was working to satisfaction.

4.3.2 Cultivation in AB10 Medium with 10 mM Citrate*

To increase overall growth the citrate concentration was raised from 4 to 10 mM. Temperature, DO and pH were not changed. A concentrated seed culture and overgrown membrane filters were used for inoculation. The seed culture was again injected via a syringe to a starting OD of 0.1. The filter membranes were pretreated as described in Chapter 3.1.3, placed on agar plates and cultivated for 48 hours at 30 °C. Three filters were used for inoculation and arranged in series at the inner wall of the glas elements of the BTR.

Fig. 4.6 shows the growth, temperature, DO and pH over time. Again, the first two days were operated as batch. After the first day the growth leveled off and was limited by substrate availability.

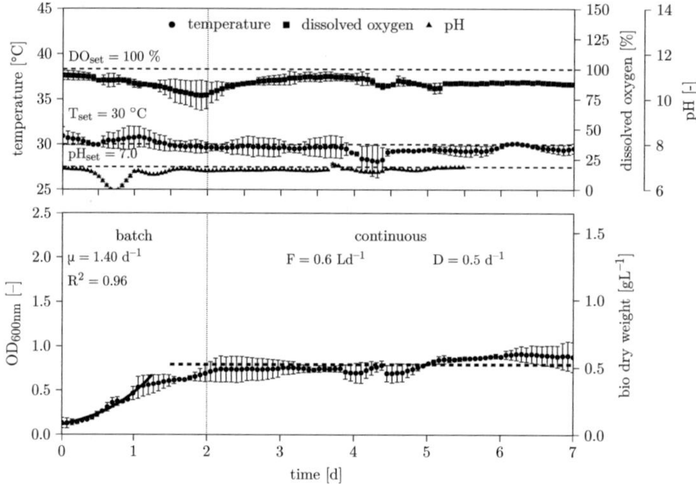

Figure 4.6: Cultivation experiment of *P. putida* KT2440 in the BTR on AB10 medium with 10 mM citrate as carbon source. The first 2 days were run as batch, followed by a continuous cultivation ($D = 0.5\ d^{-1}$).

---

* Data was partially obtained in cooperation with Yvonne Kraus (2013) as part of her bachelor thesis, TU Braunschweig.

The growth rate during the first day was given with $\mu = 1.40$ d^{-1} and was again considerably smaller than measured in the shaken flask experiments with $\mu = 11.3$ d^{-1} (compare Chapter 4.1). However, the reached maximum $OD_{max} = 0.80$ of both cultivation types, shaken flask and BTR, were similar.

The temperature of 30 °C was constant during the experiment. DO levels, however, stayed again below the targeted 100 %, reaching an average 80 %. The used DO sensor resulted in smaller errors during the experiment. pH control worked up to day 6, when data acquisition stopped. Nevertheless, the control as such continued till the end of the experiment. During the exponential growth phase a decrease in pH to 6 was perceptible. This was a consequence of the produced organic acids (e.g., acetic, lactic and pyruvic acid) which were found in some samples and measured via HPLC. pH control was able to return the pH to 7. As a result, time intervals of the PD controller were decreased to dt $= 10$ s and Pt $= 25$ s, respectively. With a specific growth rate of $\mu = 1.40$ d^{-1}, which was higher than the dilution rate of D $= 0.5$ d^{-1} no biomass washout took place and the slight increase in OD during the continuous cultivation phase occurred due to substrate limited growth. These experiments resulted in a slight increase in wall growth, but biofilm amounts were not adequate for further analysis. The membrane filters were covered with a weak biofilm matrix, only.

4.3.3 Cultivation in AB10 Medium with 10 mM Glucose*

In Fig. 4.7 the growth of *P. putida* KT2440 on a modified AB10 medium with 10mM glucose as carbon source instead of citrate is depicted. Temperature, DO and pH were not changed. No seed culture was used as inoculation but overgrown membrane filters and in gellan gum immobilized *P. putida* KT2440 cells were placed inside the biofilm tube reactor as initial biomass. The filter membranes were treated as previously described. The immobilization of the bacterial cells in the hydrogel is described in Chapter 3.1.4. The batch cultivation duration was reduced to 1.25 d as a result of the previously discussed cultivation on AB10 with 10 mM citrate (compare Chapter 4.3.2). The specific growth rate of $\mu = 3.45$ d^{-1} doubled, compared to the growth on citrate. The maximum OD was $OD_{max} = 0.67$ and thus a little below the values for the 10 mM citrate cultivation, with $OD_{max} = 0.80$. The temperature was again constant throughout the experiment.

---

* Data was partially obtained in cooperation with Vincent Wiegmann (2012) and Yvonne Kraus (2013) as part of their bachelor theses, TU Braunschweig.

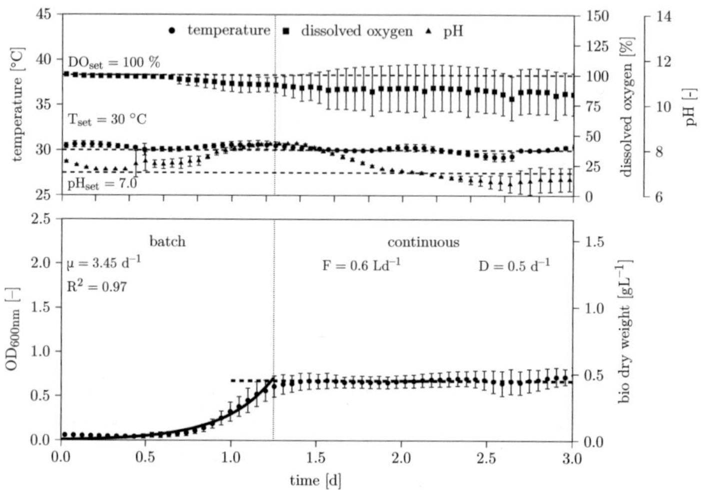

Figure 4.7: Cultivation experiment of *P. putida* KT2440 in the BTR on AB10 medium with 10 mM glucose as carbon source. The first 1.25 days were run as batch, followed by a continuous cultivation (D = 0.5 d^{-1}).

DO levels started at 100 % decreased over time to an average of 80 %. The changed PD controller settings resulted in an overshoot in pH regulation during growth. The maximum pH reached was above 8, thus as consequence the PD settings were tuned again (pH$_{min}$ = 6.95, pH$_{max}$ = 7.05, pH$_{hy}$ = 0.01, dp = 1, dt = 10 s and Pt = 10 s). After three days of cultivation experiments were stopped since the biofilm build-up on the iron netting partially detached. Similar detachment processes are discussed in Chapter 4.3.6.

To evaluate the changed PD settings and the influence of limited oxygen levels further growth experiments on AB10 with 10 mM glucose were needed. Fig. 4.8 shows the results of *P. putida* KT2440 on a modified AB10 medium with 10 mM glucose as carbon source. The data for the first 3.5 days of cultivation, with respect to OD, temperature and DO was the average of triplicates. Continuous pH values and the on going cultivation data were obtained from a single experiment. A seed culture was injected via a syringe to a starting OD of 0.1. The batch phase ended after 18 h of cultivation and the continuous phase (D = 0.5 d^{-1}) was started.

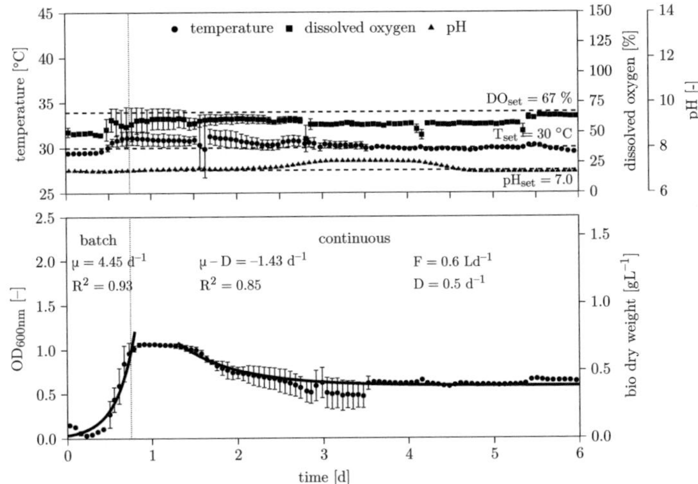

Figure 4.8: Cultivation experiment of *P. putida* KT2440 in the BTR on AB10 medium with 10 mM glucose as carbon source. The first 18 hour of cultivation were operated as batch, followed by a continuous phase (D = 0.5 d^{-1}). Targeted oxygen levels were close to 67 % saturation.

The increased maximum OD of $OD_{max} = 1.10$ was probably caused by a higher and more vital initial biomass compared to inoculation via filter membrane or cells immobilized in a gellan hydrogel as discussed in the previous Chapter 4.3.2.

This also explains a slightly increased specific growth rate of $\mu = 4.45$ d^{-1} since the bacterial cells grown on the membrane had limited access to substrate and water, while the cells in the hydrogel were exposed to high temperatures during the immobilization process. Interestingly, after 1.5 days of cultivation wash out or cell death was observed. The specific growth rate dropped below the current dilution rate. After three days of cultivation the OD leveled off to 0.65 and stayed constant till the end of the experiment. This is comparable to the biomass concentration in the previous experiment at similar conditions. The substrate limited growth resulted in constant growth after day 3, otherwise oxygen level, pH values or temperature would have changed, but data showed no major variations. The decrease in biomass could also be explained by the increased viability of the seed culture. Pre-grown on LB medium, the cells were able to store important metabolites, allowing faster growth during the batch phase, rapidly decreasing when substrate availability ended but still allowing to grow at an identical rate as the dilution rate. Finally, stored metabolites were fully catabolized and the specific growth rate dropped. The changed PD controller settings resulted in an improved pH regulation. The increase after 2.5 days of cultivation was most probably caused by a failing peristaltic pump, which was not recognized fast enough. The changed DO level to 67 % did not have an identifiable influence on the overall growth.

4.3.4 Cultivation in LB Medium

Fig. 4.9 shows the result of *P. putida* KT2440 growth on a modified LB complex medium. The concentrations of trypton and yeast extract were reduced to 1 gL^{-1} and 0.5 gL^{-1}, respectively, to limit the maximal reachable OD and to allow image analysis. Temperature and pH were not changed, targeted DO was set back to 100 %. Only overgrown membrane filters were used for inoculation. The cultivation was again divided into a batch phase of 18 h, followed by the continuous operation (D = 0.5 d^{-1}).

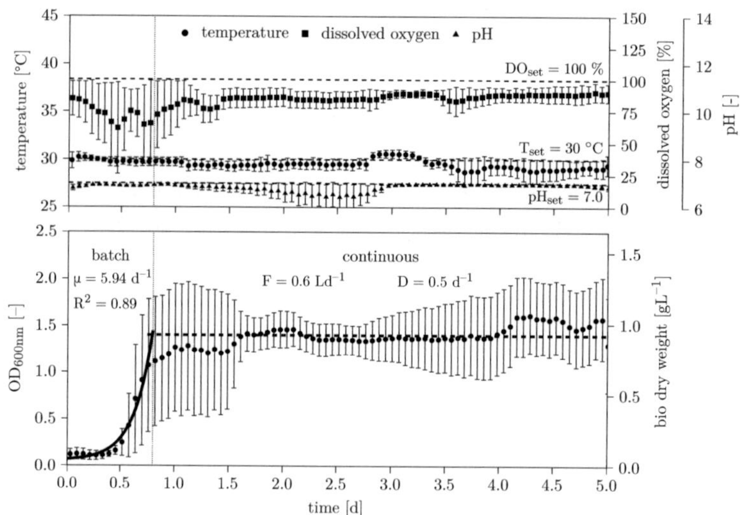

Figure 4.9: Cultivation experiment of *P. putida* KT2440 in the BTR on modified LB medium. The first 18 hours were run as batch, followed by a continuous cultivation (D = 0.5 d^{-1}).

A specific growth rate of $\mu = 5.94$ d^{-1} was reached which is only 2.4 times smaller than measured in the shaken flask culture. The continuous phase was characterized by a small increase in biomass with a maximum Optical Density of OD$_{max} = 1.40$ over cultivation time, but since the dilution rate was clearly below the maximal growth rate, it was assumed that substrate limited growth had been reached. Temperature, DO and pH monitoring worked flawlessly. pH regulation still tended to over- or undershoot but was in general acceptable. DO level decreased during the exponential growth phase to 50 % but was not limiting. Again, the 100 % saturation was not reached, probably a result of the construction and a small gassing chamber, which, however, allowed to keep the total volume small. Thus, the oxygen transfer rate was not high enough to reach saturation during cultivation. The faster growth and higher OD levels increased the influence of small unavoidable changes on the cultivation experiment and thus on its development, reducing its reproducibility which can be seen in the significant error bars.

4.3.5 Concluding Comparison of Growth Media

Generally speaking, the changes in the media composition resulted in enhanced planktonic growth and in higher growth rates of *P. putida* KT2440 with increasing substrate and medium complexity. However, this could not be correlated to the sessile growth. Too low substrate concentrations revealed no visible biofilm growth, even after a week of cultivation. Furthermore, there was neither an increase in biofilm growth at substrate concentrations of 10 mM citrate or glucose nor on the complex LB medium. The different inoculations resulted in growth on the targeted surfaces (compare Chapter 4.3.6). Table 4.2 provides a summery of the results. It was apparent that an inoculation with a planktonic seed culture resulted in a shorter lag- and batch-phase in comparison to overgrown filter membranes or immobilized cells but did not influence final biomass concentration or general growth characteristics. The high biomass density in LB medium prevented an adequate image acquisition during cultivation.

Table 4.2: Summary of cultivations in BTR on different media and alternating inoculation types.

Cultivation medium	μ [d^{-1}]	OD$_{max}$	Batch duration [d]	Inoculation
4 mM citrate	1.18	0.50	2.00	seed culture
10 mM citrate	1.40	0.80	2.00	seed culture & overgrown filter
10 mM glucose	3.45	0.67	1.25	immobilized cells & overgrown filter
10 mM glucose	4.45	1.10 / 0.65*	0.75	seed culture
yeast & trypton	5.94	1.40	0.75	overgrown filter

* after wash out, OD leveled off to 0.65

4.3.6 Concluding Comparison of Biofilm Growth

The different surfaces targeted for biofilm growth are shown in Fig. 4.10. The diameter of the tubing was kept constant at 26 mm. Flow velocity, temperature, pH or other culture conditions were not changed unless noted. The pump was kept at 20 % power, resulting in a flow rate of 1.5 Lmin^{-1}, thus the Reynolds Number was $\sim$ 1500.

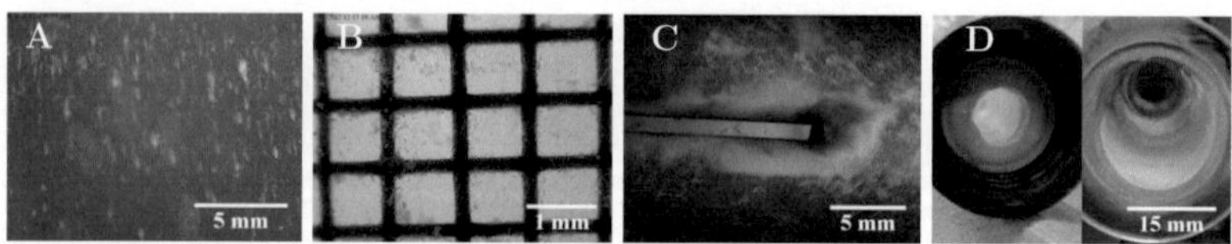

Figure 4.10: Biofilm growth on glass tube wall (A), plastic netting (B), glass object slide (C) and filter membrane (D)*.

The biofilm wall growth on etched and normal glass was poor (Fig. 4.10A) but overall biofilm built up on iron or plastic nettings (Fig. 4.10B) was promising. This was due to the changed flow pattern caused by the netting, resulting in low velocity and no flow regions, hence promoting adhesion of bacterial cells. Fig. 4.10C shows a good biofilm development on a glass object slide, placed midway between the

---

* Image was taken by Yvonne Kraus (2013) during her bachelor thesis.

walls of the tube reactor. The surface was coated with gellan gum which included immobilized cells (compare Chapter 3.1.4). The centered object slide resulted in an ideal position for image analysis in combination with low planktonic and little wall growth. However, the placement of the object slide in the mid of the tube changed the hydrodynamics and introduced turbulences. It allowed an easy removal of the biofilm including the object slide from the reactor. But further analysis with a rheometer was impossible, due to too low biomass and the destruction of the biofilm structure while scraping it off. Promising for later mechanical characterization via rheological measurements was the growth on filter membranes inlets (Fig. 4.10D). The extraction of these filter membranes often caused the demolition of the biofilm, which led to the assumption that their mechanical stability was very low. In contrast to the literature cited in Chapter 2.1 biofilm growth was not improved by limited nutrient availability. Results for AB10 medium with either citrate or glucose were comparable. General growth in LB medium was increased but did not result in an enhanced biofilm development or stability.

Fig. 4.11 presents the images taken during cultivation on AB10 with 10 mM citrate or 10 mM glucose. Fig. 4.11A to C showed little bacteria adhesion on the nettings with a small increase towards B and C as a result of the longer cultivation duration. However, the netting in Fig. 4.11D was completely covered after 9 days of cultivation. The culture conditions had not been changed but the netting had been replaced by iron.

A: Growth on *plastic* netting on AB10 medium with 10 mM citrate.

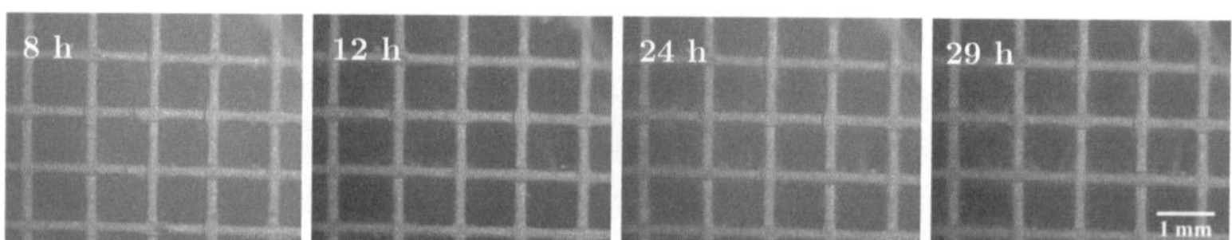

B: Growth on *plastic* netting on AB10 medium with 10 mM citrate.

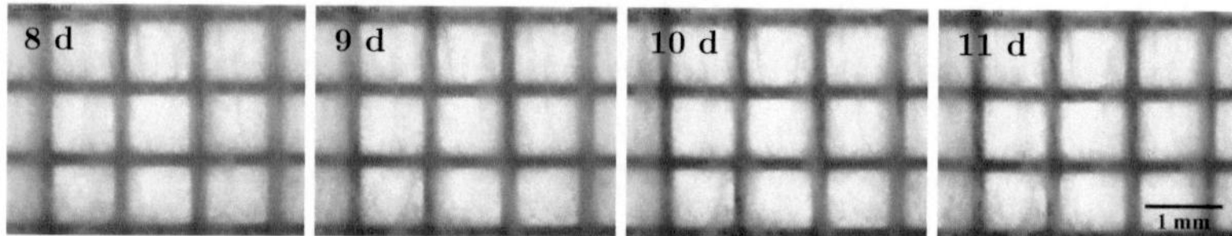

C: Growth on *plastic* netting on AB10 medium with 10 mM glucose.

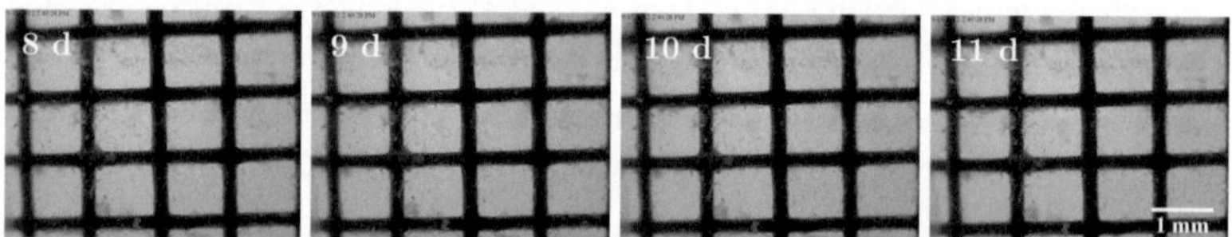

D: Growth on *iron* netting on AB10 medium with 10 mM citrate.

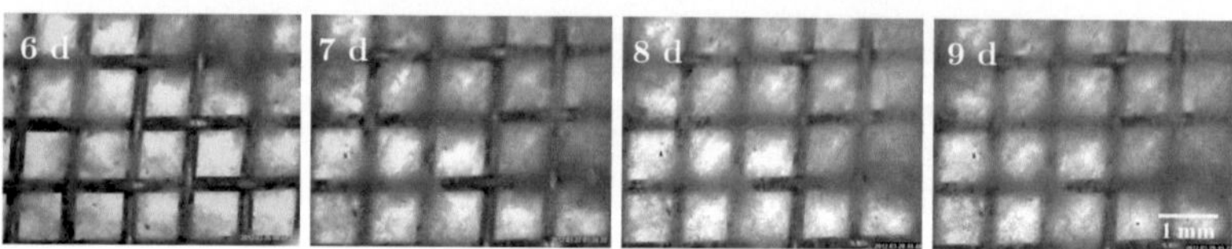

Figure 4.11: Growth of biofilm on plastic (A-C) and iron (D) nettings in AB10 medium with 10 mM citrate (A, B and D) or glucose (C). Culture conditions were kept constant.

However, the improved biofilm growth was not reproducible, repeated experiments (data not presented) showed no significant difference between an iron or plastic netting. Considering the duration of the cultivation a clear increase in biofilm density was observed over time. Nevertheless, this was expected.

Interestingly, biofilm streamers were formed and arranged in flow direction as shown in Fig. 4.11C. These were comparable to the data published by Stoodley et al. [149], who also observed an oscillation of these streamers associated to the flow velocity. Taherzadeh et al. [157] simulated this typical behavior.

Fig. 4.12 shows the cultivation on the prepared object slide with immobilized cells, which resulted in a complete coverage of the surface with a biofilm matrix after 14 days of cultivation. Problems arose when characterizing the flow pattern. The position of the object slides in the center of the tube changed the laminar flow and introduced turbulences and dead zones at the edges and at the rear of the object slides. As a consequence, the introduced stresses into the biofilm were not equally distributed and difficult to measure. This was the reason why most biofilm mass was found at the tip of the slide, which resulted in an inhomogeneous fluffy biofilm with changed density that could not be mechanically characterized with the available methods. Similar problems existed with biofilms grown on the iron and plastic nettings.

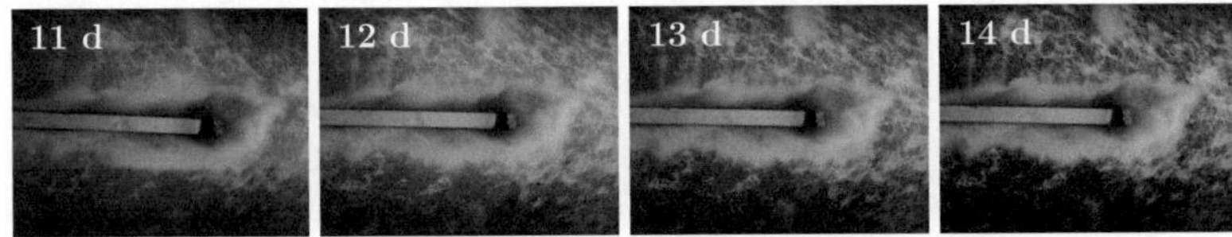

Figure 4.12: Growth of biofilm on a glass object slide in AB10 medium with 10 mM citrate (D = 0.5 d^{-1}, pH = 7, T = 30 °C, flow direction left to right).

Conclusively, biofilm growth was by chance and hardly controllable. The amount of biomass solely depended on the duration of the cultivation.

For further strength analysis of biofilms, overgrown membrane filters were used. Samples are shown in Fig. 4.13. Fig. 4.13A shows the biofilms after 48 hours of cultivation on LB agar at 30 °C. Fig. 4.13B and C present the obtained biofilm after 10 days in the BTR. While the silicon tubing (Fig. 4.13B) was completely covered by biofouling, emptying out the reactor caused massive removal of the biofilm (Fig. 4.13C). However, some filter membranes obtained good biofilm samples for future mechanical analysis (Fig. 4.13D). One major difference between the biofilm before and after the cultivation in the BTR was their water content which obviously changed their mechanical characteristics. Moreover, flattening of the membrane filter exposed them to a small strain, the inner and outer circumference of a 100 µm biofilm approximately deviated by 7.75 µm per mm biofilm length.

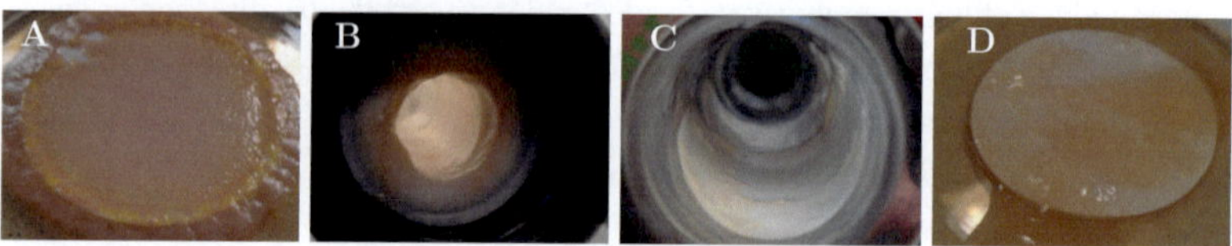

Figure 4.13: Growth of biofilm on filter membranes cultured for 48 hours on LB agar plates and then placed into the BTR for 10 days. (A) Growth on membrane filter before placement into the BTR, (B) sample of biofilm growth on the silicon tubing in comparison (C) the overgrown membrane filter after the media was removed from the BTR, (D) a good example of a biofilm cultured in the BTR.*

4.3.7 Experiments with Detachment

A further objective besides biofilm growth was to study detachment processes in biofilms. Therefore, the development of biofilms was monitored by image acquisition techniques. Fig. 4.14 shows a sample cultivation at its final stadium after approximately 11 days in total. First, the biofilm grew to cover most of the netting, followed by small detachment or some erosion, which smoothed its structure (Fig. 4.14, 126.7 to 169.4 h). The biofilm recovered, grew in thickness and eventually covered the whole area (Fig. 4.14, 169.4 to 230.8 h). Then cracks appeared which finally led to a failing overall structure. The sloughing occurred within minutes leaving some fragments behind (Fig. 4.14, 230.8 to 235.3 h), staying attached until they were also ripped off by the current (Fig. 4.14, 247.2 to 271.1 h).

This so called avalanche effect was also described by Picioreanu et al. [132] who modeled the detachment process and observed, in accordance with the experiment described, an initial erosion process. The remaining fragments were directly exposed to the fluid induced stress and failed as well [150]. At last, this process reached an end when the remaining biofilm was thin enough to withstand the shearing forces and a new biofilm growth phase began. Wagner et al. [169] found that more compact biofilm structures could be sloughed off in complete. Other possible reasons for

---

* Images were taken by Yvonne Kraus (2013) during her bachelor thesis.

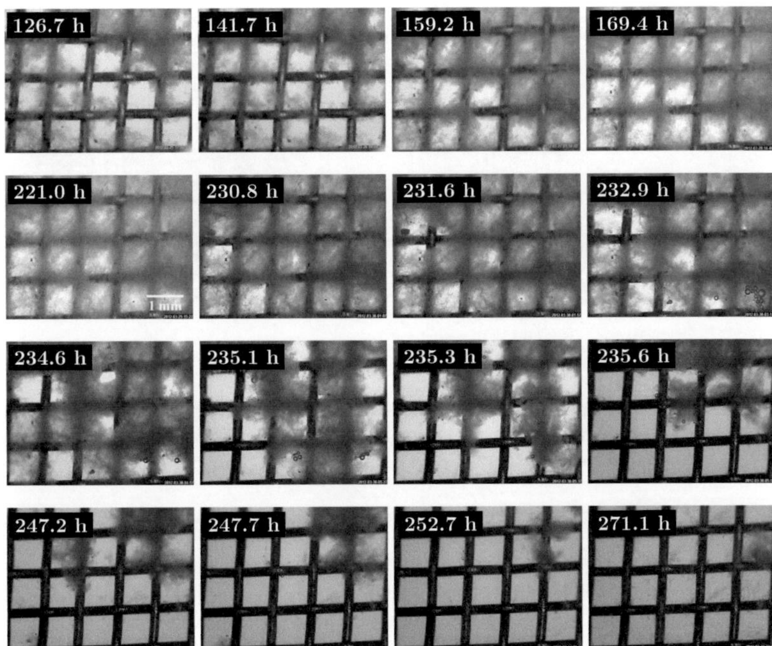

Figure 4.14: Biofilm growth and detachment during cultivation in AB10 medium with 10 mM citrate (D = 0.5 d^{-1}, pH = 7, T = 30 °C).

the sudden detachment were a critical thickness, resulting in stresses induced by the fluid finally being higher than its tensile strength or in a weakened biofilm structure. Perhaps, this was a result of finite substrate availability, due to limited diffusion into lower biofilm layers. It is known that low biofilm layers can consist of dead and inactive cells which weaken the overall structure [65, 163].

Fig. 4.15 shows the relative biofilm growth. A value of 1 is given when the maximum biofilm coverage is reached while a value of 0 means no biofilm occurrence. The data is based on the image acquisition which is shown in Fig. 4.14. These images were transfered into binaries. The ratio of black and white pixels, with the lowest being the image at the bottom right, results in 0 relative biofilm coverage while the highest ratio results in 1 relative biofilm coverage.

This value is not equivalent to a 100 % coverage and does not include the biofilm thickness. It is a measurement to compare the development of a single biofilm over

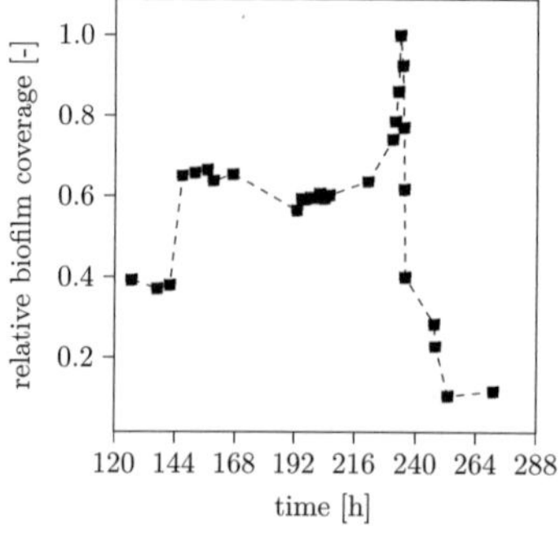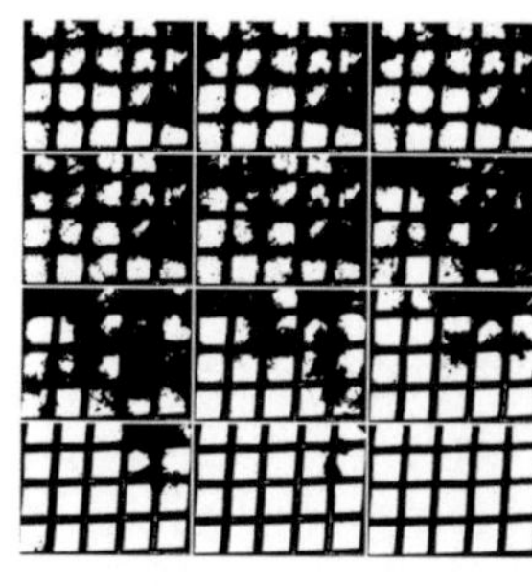

Figure 4.15: Left: Biofilm coverage of the last 6 days (126.7 to 271.1 h) of cultivation. Right: Sample binary images, the maximum coverage is defined as one, while no biofilm is set to 0 (bottom right image).

time when camera position is not changed. In this case the development of the biofilm over the last 6 days (126.7 to 271.1 h) of cultivation was computed. The correlation with the images shown in Fig. 4.14 is acceptable. The first period was characterized by exponential biofilm growth, which was followed by a period with no major changes. A light reduction in the covered surface was observed as cause of erosion processes. This period was again followed by exponential growth and interrupted by a sudden sloughing process. The biofilm was not able to recover and the already described avalanche effect occurred, slowed down but did not stop until almost complete removal of the biofilm.

4.3.8 Biofilm Mechanics*

First rheological biofilm measurements were performed to define the range for the storage and loss moduli of biofilms for the model, since data in literature varies by several magnitudes depending on microorganisms and measurement techniques as shown by Böl et al. [12]. *P. putida* KT2440 and *E. coli* K12 biofilms were cultured on the filter membranes as described in Chapter 3.1.3.

Fig. 4.16 displays the average of storage (G') and loss (G'') moduli versus frequency (ω) of multiple frequency sweeps at an oscillatory strain of $\pm 5 \cdot 10^{-4}$. The frequency was kept in the range of 0.1 to 10 s^{-1}. Error bars show the 95 % confidence interval.

* Data was partially obtained in cooperation with Steffi Günther, Institute for Particle Technology, TU Braunschweig.

The storage moduli for both species slightly increased over the measured range. G' for *E. coli* K12 was always above the one for *P. putida* KT2440. Similar results were obtained for G''. Again, G'' for *E. coli* K12 was higher than G'' for *P. putida*. In general, the changes with increasing frequency for both types of biofilms were small. G' was within 6 to 14 kPa and G'' ranged from 0.9 to 1.5 kPa.

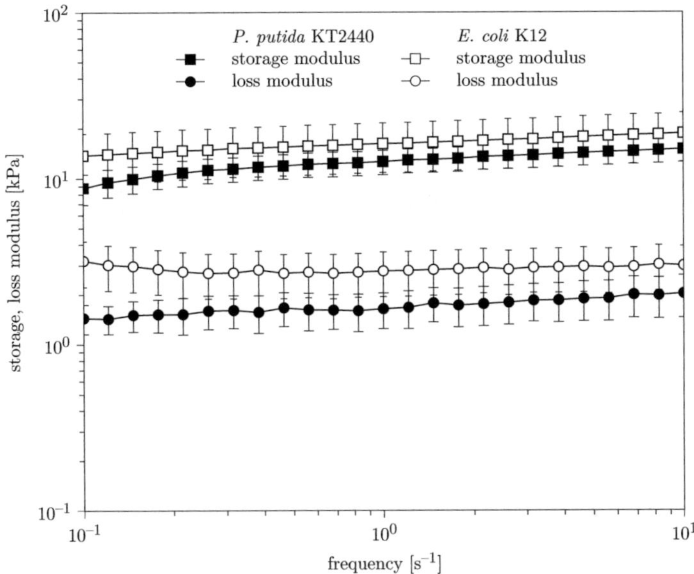

Figure 4.16: Frequency sweep of *P. putida* KT2440 and *E. coli* K12 biofilm cultured on filter membranes at an oscillatory strain of $\pm 5 \cdot 10^{-4}$. Data was averaged over 14 runs, error bars show the 95 % confidence interval.

The next step was to adapt the worm-like-chain- or the Maxwell-Wiechert-model for polymers as shown by Ehret et al. [41] and as described in Chapter 2.3.2 by Eqs. (2.48) to (2.50). Hence, the spring constants G_i and the relaxation times τ_i of each Maxwell element were computed. The relaxation modulus was fitted to Eq. (2.48) and the storage and loss moduli obtained from the frequency sweep were fitted to Eqs. (2.49) and (2.50). The number of Maxwell elements was increased from one to four.

For the frequency sweeps *P. putida* and *E. coli* biofilms grown on membrane filters were used. The results of the frequency sweep experiments for the storage and loss moduli are shown in Fig. 4.17. A model with a single Maxwell element (n = 1, dotted) was neither able to simulate the storage nor the loss moduli adequately. It was not possible to correlate both, storage and loss moduli, to the data points without large divergences. The correlation coefficient was estimated with $R^2 = 0.387$. Correlating the data to two Maxwell elements (n = 2, dashed) resulted in a good estimate of the storage but again in a poor agreement between the measured and computed loss moduli. The averaged regression coefficient was given with $R^2 = 0.407$.

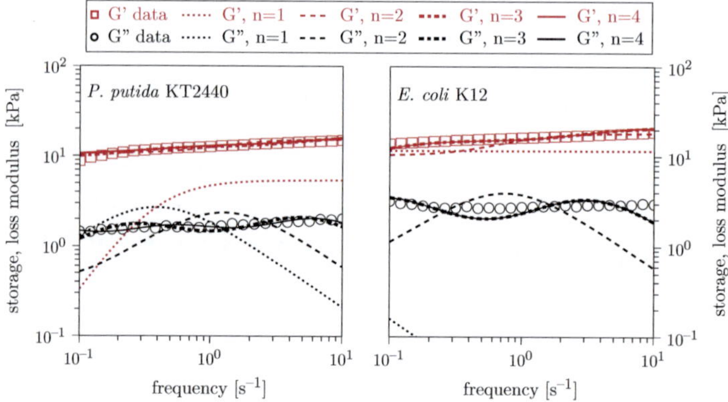

Figure 4.17: Frequency sweep experiments of filter grown biofilms of *P. putida* KT2440 (left) and *E. coli* K12 (right). Oscillatory strain was kept between $\pm\, 5 \cdot 10^{-4}$. Data was fitted to Eqs. (2.49) and (2.50) with n = 1...4. Curves for n = 3 and n = 4 for *E. coli* overlap.

Thus, this was still insufficient to describe the curves of the loss moduli for both biofilms obtained from the frequency sweep data. Three Maxwell elements (n = 3, dash dotted) finally resulted in an acceptable description of the experimental data with $R^2 = 0.836$ and four Maxwell elements (n = 4, solid line) slightly increased the model discrimination for the *P. putida* KT2440 biofilms. The curves for the storage and loss moduli for *E. coli* K12 overlap for three and four Maxwell elements, thus no improvement was observed. The overall regression coefficient was computed with $R^2 = 0.869$.

Fig. 4.18 shows the measured data of the relaxation experiments on *E. coli* K12 filter grown biofilms and the model results. Again an obvious improvement in the model discrimination with an increase in Maxwell elements was observed. The correlation coefficient increased from $R^2 = 0.72$ to 0.96, the latter was identical for three and four Maxwell elements. This was also observed in Fig. 4.17, where n = 3 and n = 4 resulted in the same model description. Conclusively, an increase in Maxwell elements improved the model discrimination as anticipated. Three Maxwell elements resulted in an overall good estimation, with a slight increase in accuracy for four Maxwell elements, which also led to a good agreement with the worm-like-chain-model approach by Ehret et al. [41] and is physically justified by the different types of attraction, which are electrostatic and dispersion forces, hydrogen bonds and entanglements (compare Chapter 2.3.2).

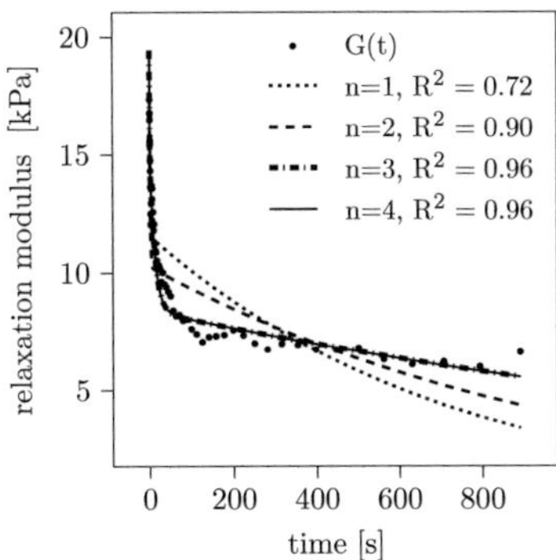

Figure 4.18: Results of relaxation experiments on *E. coli* K12 filter grown biofilms. The strain was kept constant at $5 \cdot 10^{-4}$. Data was fitted to Eq. (2.48).

The spring constants (G_i) and relaxation time constants (τ_i) obtained by fitting the storage, loss and relaxation moduli to Eqs. (2.48) to (2.50) with four Maxwell elements (n = 4) are summarized in Table 4.3. A direct comparison of the spring constants G_i and the relaxation time constants τ_i for *P. putida* KT2440 with data published by Ehret et al. [41] for *P. aeruginosa* FRD1 yielded, that the values were within the same range with respect to the relaxation time constants but one magnitude higher for the spring constants. The theoretical approach suggested that the spring constants for the same attraction forces did not change, thus proposing that the difference in the *P. putida* KT2440 and *P. aeruginosa* FRD1 EPS-matrix was small. This idea was strengthened by the fact that the obtained data for the *E. coli* KT12 varied by several magnitudes with respect to the spring constants G_i, suggesting a different EPS composition based on other polysaccharides [9]. A possible reason for the rise in the spring constants was found in the increased concentration of divalent ions, especially 20 mM Mg^{2+} compared to 10 mM Ca^{2+} for the published data. The influence of Ca^{+2} and Mg^{2+} are comparable since they are both divalent ions and play a major role in the regulation of polysaccharide synthesis and cell attachment [2, 181]. The influence of Ca^{2+}, however, is considered higher, thus the increased spring constants can be explained by either the concentration or as a result of a different organism, and hence a different EPS.

Table 4.3: Fitting parameters of the worm-like-chain- or Maxwell-Wiechert-model for four Maxwell elements. Data fitted to the storage, loss and relaxation moduli from *P. putida* KT2440 and *E. coli* K12 filter membrane grown biofilms. Data for *P. aeruginosa* FRD1 was taken from Ehret et al. [41].

Maxwell element #i	*P. putida* KT2440 G_i [kPa]	τ_i [s]	*E. coli* K12 G_i [kPa]	τ_i [s]	*P. aeruginosa* FRD1 G_i [kPa]	τ_i [s]
1	3.667	0.146	6.329	0.314	0.170	0.0607
2	2.439	1.909	6.633	11.873	0.361	1.1088
3	2.708	24.692	7.874	1993.1	0.349	16.009
4	8.233	2311.3	0.528	160431	1.332	1925.4

4.4 Biomimetic Gellan Based Hydrogels

A biofilm is a very complicate system and influenced by many different variables and environmental aspects as discussed in detail in Chapter 2.1. As a result of this and because the problems that arose while cultivating biofilms for mechanical characterization a simplified physico-chemical model was required. This ought to be able to mimic the biofilm mechanics and reduce its complexity. The material of choice was gellan gum. Gellan has a thermoreversible sol-gel transition and the viscoelastic properties can be influenced by mono- and divalent ions as discussed in Chapter 2.2.2. The influence of Na^+ and K^+ on the sol-gel temperature T_{SG}, storage (G') and loss (G'') moduli was studied by Miyoshi et al. [101, 102] and the influence of Mg^{2+} by Sworn et al. [156] and Izumi et al. [68], respectively. T_{SG} is in general higher for Ca^{2+} than for Mg^{2+}, while the gellation concentration for Na^+ is lower than for K^+. Na^+ and Mg^{2+} were chosen as ions to control the viscoelasticity of the gellan gum based biomimetic hydrogels. These were mechanically characterized via rheometer techniques and modeled as response surfaces, a design of experiments (compare Chapters 2.4 and 3.3). The test methods of the mechanical characterization were either stress relaxation, creep or dynamic mechanical. The results are presented in the following sections.

4.4.1 Defining the Design Space

The gellan manual [18] states a maximum gel hardness for hydrogels containing 1 %(w/v) gellan and between 6 to 8 mM Mg^{2+}. Mg^{2+} results at a concentration above 8.4 mM in a gelation temperature above 50 °C . A sol-gel transition temperature below 50 °C is reached for concentrations with less than 100 mM Na^+ or K^+, respectively. Own experiments (not presented) showed that concentrations below 0.5 %(w/v) gellan, 30 mM Na^+ and 5 mM Mg^{2+} resulted in weak still highly viscous gels which were not suitable to simulate biofilm properties. Thus, the center point (CP) was defined as 0.75 %(w/v) gellan, 40 mM Na^+ and 6.7 mM Mg^{2+}, being in between the above values. The corners of the cube represent the minimum and maximum concentrations of the values stated above and theses agree with most of the data of the cited authors which are summarized in Table 2.1. The chosen ions and their concentrations, however, deviated from published data. The fully factorial design included 6 replicates of the CP, all other 14 hydrogel compositions were measured in triplicates. Table 4.4 summarizes the concentrations of each experiment, which reach from 0.33 to 1.17 %(w/v) for gellan, 3.9 to 9.5 mM for Mg^{2+} and 23 to 57 mM for Na^+. The frequency dependence was also included and varied from 0.1 to 10 s^{-1} over 25 levels due to its influence on G' and G''.

Table 4.4: Central composite design with concentrations for gellan, Mg^{2+} and Na^+ and the coded values ($-\alpha$, -1, 0, $+1$, $+\alpha$). Hydrogel #1 to #8 are the corners of the cube, #9 to #14 are the outer points of the star, and #15 is the CP, compare Chapter 2.4.

Hydrogel #	x_{Gellan} [%](w/v)		$c_{Mg^{2+}}$ [mM]		c_{Na^+} [mM]	
1	0.50	-	5.0	-	30	-
2	0.50	-	8.4	+	30	-
3	0.50	-	5.0	-	50	+
4	0.50	-	8.4	+	50	+
5	1.00	+	5.0	-	30	-
6	1.00	+	8.4	+	30	-
7	1.00	+	5.0	-	50	+
8	1.00	+	8.4	+	50	+
9	0.33	$-\alpha$	6.7	0	40	0
10	1.17	$+\alpha$	6.7	0	40	0
11	0.75	0	6.7	0	23	$-\alpha$
12	0.75	0	6.7	0	57	$+\alpha$
13	0.75	0	3.8	$-\alpha$	40	0
14	0.75	0	9.5	$+\alpha$	40	0
15	0.75	0	6.7	0	40	0

4.4.2 Amplitude Sweeps*

The first measurements were performed to define the region, where the hydrogel behaves approximately linear-viscoelastic and comparable to the filter grown biofilms. Figs. 4.19 and 4.20 show the G' and G'' as well as the phase angle of two hydrogels (#9 and #10) versus the strain and the filter grown biofilms of *P. putida* KT2440, respectively. The applied strain γ was increased from 10^{-4} to 1 at a constant frequency of $1\ s^{-1}$. Between 10^{-4} and 10^{-3}, G' and G'' were approximately constant. The curve shapes for G' and G'' as well as for the phase angle for both gels and the biofilm are clearly visible. They represent the data of the lowest and highest concentrated gellan hydrogels, which contained 0.33 %(w/v) and 1.17 %(w/v) gellan and a 48 hours old bacterial lawn. The results are comparable to those of García et al. [51] for a 0.1 %(w/v) gellan gel with $c_{Na^+} = 220$ mM. In comparison to the biofilm data, a similar development in terms of G' and G'' as well as for the phase angle was determined.

---

* Data was partially obtained in cooperation with Steffi Günther, Institute for Particle Technology, TU Braunschweig.

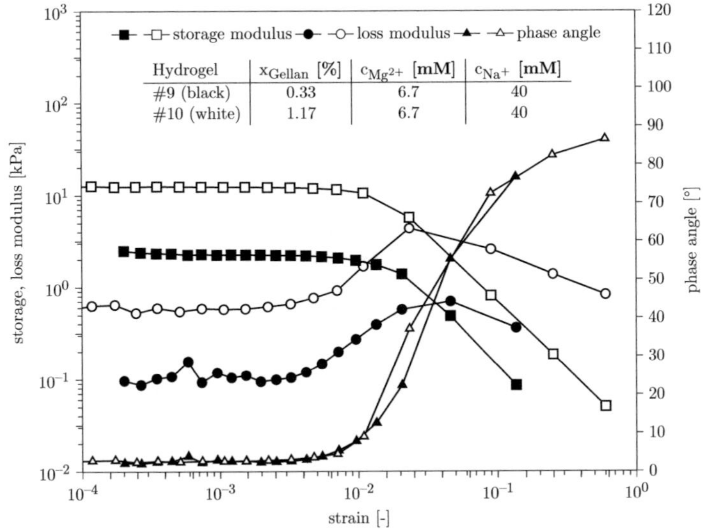

Figure 4.19: Amplitude sweep of two different gellan-based hydrogels, both only differ in their gellan content, $x_{Gellan/9} = 0.33$ %(w/v) (black) and $x_{Gellan/10} = 1.17$ %(w/v) (white) with a constant ion concentration of $M_{Mg^{2+}} = 6.7$ mM and $M_{Na^+} = 40$ mM. The frequency was set constant to 1 s^{-1}.

Te Nijenhus [161] described the straight line at lower strains as plateau, which can also be observed in agarose or x-carrageenan-based systems containing similar salts. For strains (γ) between 10^{-4} and $5 \cdot 10^{-3}$ the phase angle was small, pointing to a behavior similar to that exhibited by linear-elastic or Hookean material. For larger deformations, the phase angle increased to 90°, this changed the behavior of the material from Hookean solid-like behavior ($\delta = 0°$) to Newtonian liquid-like behavior ($\delta = 90°$), compare Eq. (2.38). However, when leaving the constant region, the error increased, thus reducing the accuracy. The linear range did not change between the two hydrogels, and the tendency towards an increase in the phase angle and a transition from a Hookean solid to a Newtonian liquid was similar and comparable to the biofilm data. Hence, all further frequency sweep experiments were performed at an oscillatory strain γ of $\pm 5 \cdot 10^{-4}$ to ensure reproducibility.

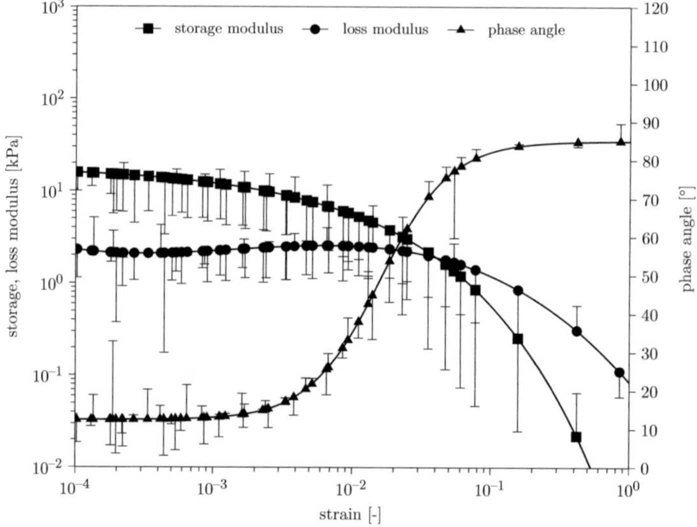

Figure 4.20: Amplitude sweep of filter grown biofilms of *P. putida* KT2440. The frequency was set constant to 1 s^{-1}.

4.4.3 Frequency Sweeps*

The first modeling approaches ignored the influence of frequency on G$'$ and G$''$, leading to an unsatisfactory model estimate. However, according to te Nijenhuis [161], a gel has a depicted maximum which is followed by a minimum in G$''$, while G$'$ yields a straight line when plotted versus the frequency in a double logarithmic graph (compare Fig. 4.21). Due to the general limitations of the measurement techniques, it was difficult to reach the maximum at very low frequencies [161].

Fig. 4.21 shows G$'$ and G$''$ versus a frequency ranging from 0.1 to 10 s^{-1} at an oscillatory strain γ of $\pm$ 5 $\cdot$ 10^{-4}. A constant G$'$ or slight increase in G$'$ and a minimum for G$''$ were observed. G$''$ was always considerably smaller than G$'$ due to the low bonding energies between the molecules [161], which is according to Mayer et al. [98] also typical for biofilms.

G$'$ was approximated by linear regression, but a quadratic or cubic polynomial was more applicable for G$''$. Higher concentrations of salt or polysaccharides resulted

---

* Data was partially obtained in cooperation with Steffi Günther, Institute for Particle Technology, TU Braunschweig.

in smaller changes in G' or G'' with increasing frequency. Gels with a high x_{Gellan} (e.g., #10) had a smaller slope for G' and a less pronounced curvature in G'' than gels with smaller amounts of x_{Gellan}. The slope of G' varied from 0.015 to 0.13 depending on the hydrogel and was smaller but within the range of the value of 0.20 established by te Nijenhuis [161]. These results were within the same dimension as the data of previous studies of similar concentrated gellan systems or comparable gels in an identical frequency range [100, 102, 121, 122, 127, 129]. Confirming the results reported by te Nijenhuis [161], Fig. 4.21 shows measurements of the gel-like region of the hydrogel, which is in between the liquid-like and the rubber-like region and is defined by a straight line in a double-logarithmic plot of G' versus frequency. However, the gellan-based hydrogels with high gellan content and added salts displayed only weak gel properties, consistent with the results of Cesàro et al. [20], te Nijenhuis [161], and Miyoshi et al. [100]. Despite the small variation, the frequency dependence was important as an additional logarithmic factor $f = \log(\omega)$ in the CCD and increased the quality of the estimated response surfaces.

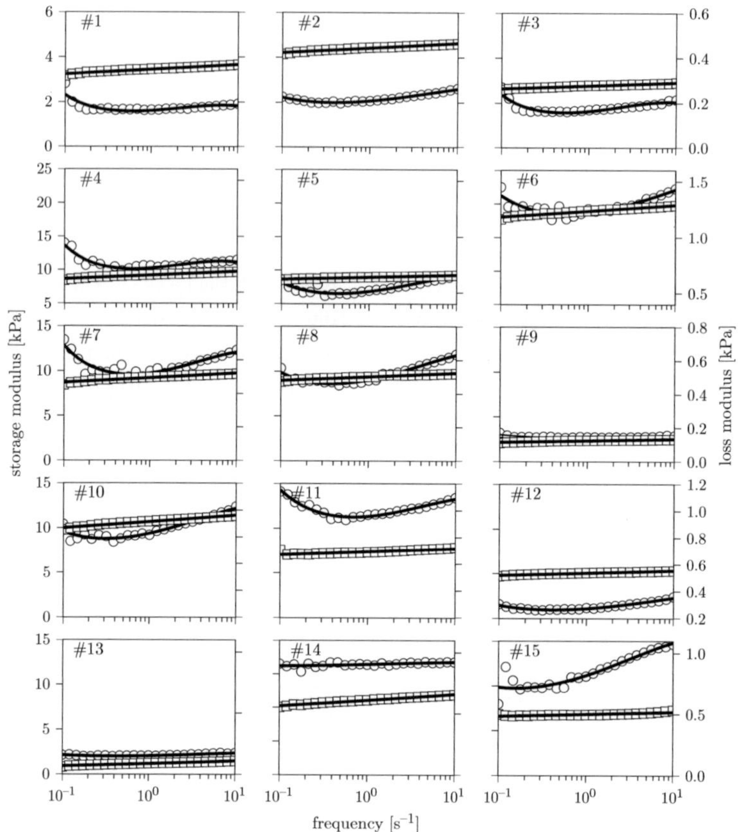

Figure 4.21: Storage (G′, square) and loss (G″, circle) moduli versus the frequency of the CCD data. Experimental mean values are presented as symbols, while the lines are the fitted polynomials. All measurements were performed in triplicate. G′ shows a linear tendency, while G″ can be described by a quadratic or cubic polynomial.

4.4.4 Surface Response Methodology

The presented response surfaces were the results of fitting the data of 15 hydrogels with different composition (#1 to #15, compare Table 4.4) to Eq. (3.4) with R being G' and G'', respectively. Each model term was tested to improve the overall model and chosen if statically significant, only. The model residuals were checked whether normally distributed or otherwise a power transformation factor λ was computed. λ was used to power transform G' and G'' and the results were again fitted to Eq. (3.4). If λ was close to one, the model was considered to include all effects and the iteration process was stopped. The solution is presented below. The responses were power transformed with $\lambda = 0.5 \rightarrow R' = \sqrt{R}$ for G' and $\lambda = 0 \rightarrow R' = \log(R)$ for G''. This resulted in a suggested model for G' with $F = 504.60$ and $R^2 = 0.88$. There was only a 0.01 % chance that a F-value that high was due to noise. The adequate precision ratio was 75.19 which strengthened the above results. This, finally led to the following significant model terms, x_{Gellan}, $c_{Mg^{2+}}$, c_{Na^+}, f and $x_{Gellan}c_{Na^+}$:

$$\sqrt{G'} = a_0 + a_1 x_{Gellan} + a_2 c_{Mg^{2+}} + a_3 c_{Na^+} + a_4 f \\ + a_5 x_{Gellan} c_{Na^+} \; . \tag{4.1}$$

The recommended model function for G'' was given with $F = 305.44$ and $R^2 = 0.83$. Although these values were lower than those for G', it still resulted in a 0.01 % chance being caused by noise, only. The adequate precision ratio was given with 61.99, thus resulting again in an adequate model description. A probability of 0.95 proved that the model terms x_{Gellan}, $c_{Mg^{2+}}$, c_{Na^+}, f, $x_{Gellan}c_{Na^+}$, x_{Gellan}^2, $c_{Mg^{2+}}^2$ and f^2 were significant:

$$\log(G'') = a_0 + a_1 x_{Gellan} + a_2 c_{Mg^{2+}} + a_3 c_{Na^+} + a_4 f \\ + a_6 x_{Gellan} c_{Na^+} \\ + a_{11} x_{Gellan}^2 + a_{12} c_{Mg^{2+}}^2 + a_{14} f^2 \; . \tag{4.2}$$

Thus, the final Eqs. (4.1) and (4.2) allowed to predict G' and G'' for any hydrogel within the design space or compute the gellan content x_{Gellan}, the ion concentration $c_{Mg^{2+}}$ and c_{Na^+} at a certain frequency to give a desired G':

$$\sqrt{G'} = -161.103 + 250.869 x_{Gellan} + 9.136 c_{Mg^{2+}} \\ + 3.215 c_{Na^+} + 2.911 f - 4.484 x_{Gellan} c_{Na^+} \tag{4.3}$$

and

$$\begin{aligned}
\log(G'') = {} & -2.311 + 5.452 x_{Gellan} + 0.621 c_{Mg^{2+}} \\
& + 0.030 c_{Na^+} + 0.020 f \\
& - 0.047 x_{Gellan} c_{Na^+} \\
& - 2.008 Gellan^2 - 0.038 c_{Mg^{2+}}^{\;2} \\
& + 0.045 f^2 \; .
\end{aligned} \tag{4.4}$$

Eqs. (4.3) and (4.4) show the resulting empirical model equations for the response surface for G' and G'', respectively. The influence of the gellan and ions for G' appeared to be mostly of linear nature, while the relation for G'' was more complex. Interestingly, the influence of frequency on the model correlated with the previous observation and resulted in a linear (G') or quadratic (G'') relation. This agrees with data published by te Nijenhuis [161] and Liu et al. [92] in the frequency interval of 0.1 to 10 s^{-1}.

Figs. 4.22 and 4.23 illustrate the results of G', the model responses are shown as surfaces. In Fig. 4.22, the maximum storage modulus reached was close to 20 kPa at a maximum x_{Gellan} of 1.17 %(w/v) and a highest $c_{Mg^{2+}}$ of 11 mM. The minimum was located at the least x_{Gellan} and a minimal $c_{Mg^{2+}}$ which was below 2 kPa. The maximum and minimum values shown in Table 2.1 are close to the ones obtained for the CCD. However, the used ion concentrations were different and x_{Gellan} often higher than compared to Noda et al. [122] or Pèrez-Campos et al. [129]. This made direct comparisons difficult. At constant c_{Na^+}, the addition of Mg^{2+} or x_{Gellan} clearly increased G' and in combination the strengthening influences intensified. In Fig. 4.23 at constant $c_{Mg^{2+}}$ and low x_{Gellan}, the availability of Na^+ has a weakening or leveling effect with increasing concentration resulting in a maximum G' of 25 kPa at the lowest c_{Na^+} but highest x_{Gellan}. The higher c_{Na^+}, the smaller was the influence of the total x_{Gellan} on G'.

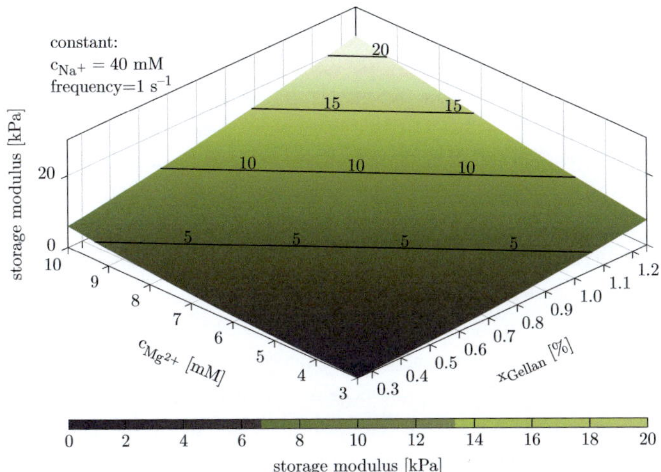

Figure 4.22: Results of fit for the storage modulus (G′). The modeled results are represented by the surface. x_{Gellan} and $c_{Mg^{2+}}$ were varied, c_{Na^+} (40 mM) and the frequency (1 s^{-1}) were kept constant.

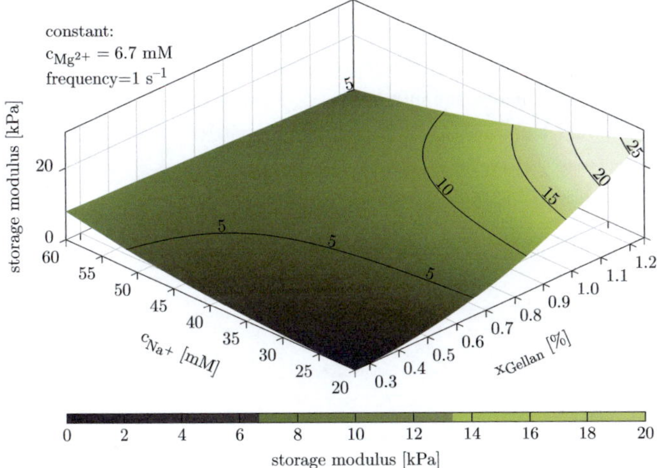

Figure 4.23: Results of fit for the storage modulus (G′). The modeled results are represented by the surface. x_{Gellan} and c_{Na^+} were varied, $c_{Mg^{2+}}$ (6.7 mM) and the frequency (1 s^{-1}) were kept constant.

In general, similar influences of $c_{Mg^{2+}}$ and x_{Gellan} were observed. If c_{Na^+} was constant an increase in either x_{Gellan} or $c_{Mg^{2+}}$ resulted in a gain in G'. However, at constant $c_{Mg^{2+}}$, high x_{Gellan} and low c_{Na^+}, high values for G' were measured but small values at low x_{Gellan}. Furthermore, at low x_{Gellan} G' rose with increasing c_{Na^+} but decreased at high x_{Gellan}. Finally, at the maximum c_{Na^+} this resulted in a 0.5 %(w/v) gellan based hydrogel with high c_{Na^+} being stiffer than a 1.0 %(w/v) gellan based hydrogel with similar c_{Na^+}. It can be summarized that x_{Gellan} as well as $c_{Mg^{2+}}$ increased G'. A gain in c_{Na^+} tended to weaken highly but stiffened poorly concentrated gellan based hydrogels.

Figs. 4.24 and 4.25 display the results for G''. The model solutions are plotted as surface. The major difference compared to the results of G' is a clear maximum in G'' when plotted against x_{Gellan} and $c_{Mg^{2+}}$. This maximum was already described by Sworn et al. [156]. Again, Na^+ as monovalent ion had a different effect on G'' than the divalent ion which was comparable to G' in Fig. 4.23. The maximum loss modulus was close to 1.5 kPa while the minimum was around 0.1 kPa. This is within the range of the values found in the cited literature summarized in Table 2.1, e.g., Noda et al. [122] and Oliveira et al. [127]. In Fig. 4.24, G'' is identical for different $c_{Mg^{2+}}$, this was caused by the quadratic relation. For a low x_{Gellan} at constant $c_{Mg^{2+}}$, differences in G'' were small, however tended to increase towards higher gellan levels. The maximum G'' was reached at $c_{Mg^{2+}}$ between 8.0 to 9.0 mM and x_{Gellan} between 0.75 to 1.0 %(w/v), respectively, and agreed with the data provided by Carl Roth [18]. Finally, the influence of c_{Na^+} on G'' was in accordance with G'. At low c_{Na^+} and high x_{Gellan} G'' reached its maximum, but decreased rapidly if Na^+ was increased. However, at low x_{Gellan} an increase in c_{Na^+} was followed by an only small gain in G'' (compare Fig. 4.25).

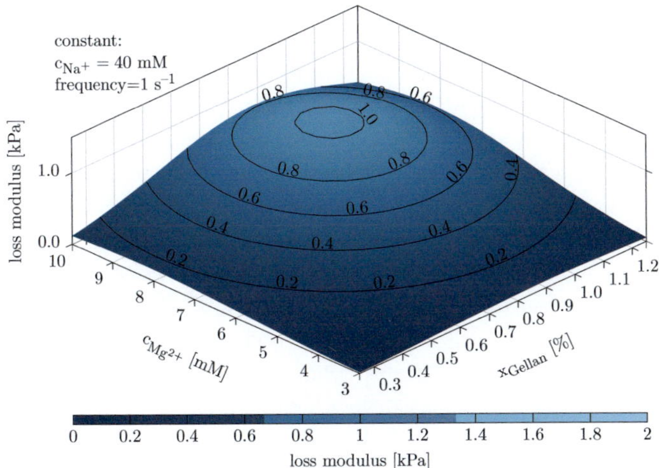

Figure 4.24: Results of the fit for the loss modulus (G''). The modeled results are represented by the surface. x$_{Gellan}$ and $c_{Mg^{2+}}$ were varied, c_{Na+} (40 mM) and the frequency (1 s^{-1}) were kept constant.

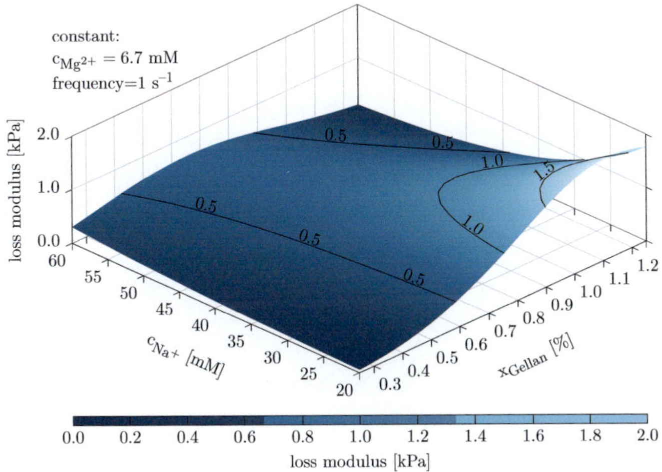

Figure 4.25: Results of the fit for the loss modulus (G''). The modeled results are represented by the surface. x$_{Gellan}$ and c_{Na+} were varied, $c_{Mg^{2+}}$ (6.7 mM) and the frequency (1 s^{-1}) were kept constant.

4.4.5 Quality of the Model

The overall quality of the models is shown in Fig. 4.26. It displays how well the models fit the measured data. The x-values represent the number of the measured or computed hydrogels (compare Table 4.4) in the frequency interval from 0.1 to 10 s^{-1}. The black lines are the results of the experiments plotted with the confidence interval of 95 % (gray), the dashed lines give the predicted or calculated values for G' and G'' based on the computed empirical models (compare Chapter 4.4.4). In general, the obtained models showed a good consistency between modeled and measured data. Exceptions were the hydrogels numbered 1, 3, 4, 7 and 13 for G' or 2, 3 and 7 for G''. These results were slightly outside the confidence interval (95 % confidence interval) but still close to the measured values. The given $R^2 = 0.88$ and $R^2 = 0.83$ are reasonable for such an empirical physico-chemical model.

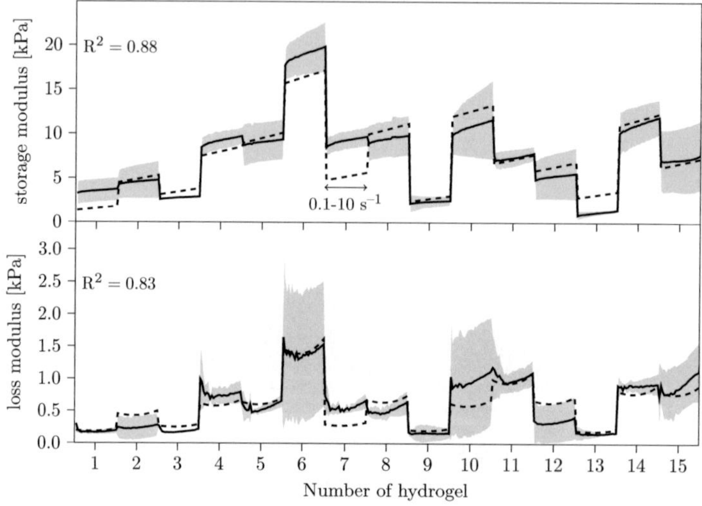

Figure 4.26: Quality of the model data (dashed) versus experimental data (solid) with 95 % confidence interval (gray).

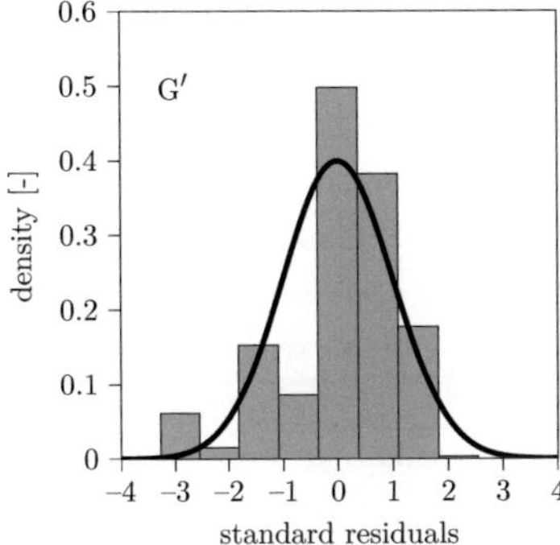
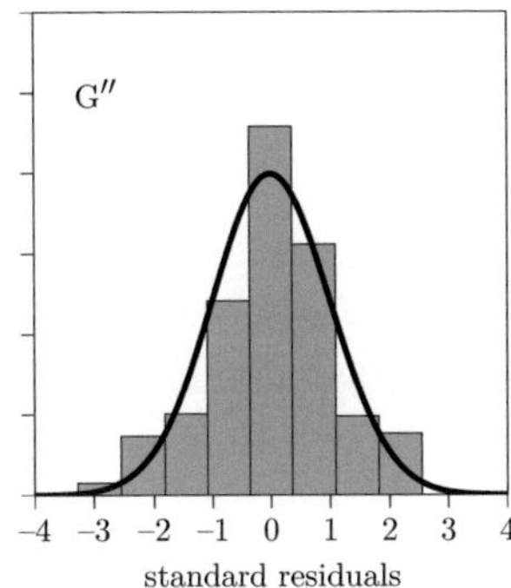

Figure 4.27: Gaussian distribution of standard residuals for the storage and loss moduli. The normal curves are shown in black, the density of the standard residuals of the models are shown as gray bars.

Fig. 4.27 presents the Gaussian distribution of standard residuals for the storage and loss moduli. The normal curves are black and the density of standard residuals of the models are shaded gray. Based on the assumption that non-systematic errors are random and normally distributed, the deviation from the ideal curve (black) is a tool to decide weather a model represents the data accurately and includes all effects. If the residuals were not normally distributed this would imply an error in the mathematic model or a systematic error in the measurements [7, 144]. However, the residuals for the storage and loss moduli are clearly normally distributed.

4.4.6 Relaxation of Gellan*

Fig. 4.28 shows a stress relaxation experiment for the center point (CP) hydrogel #15 (compare Table 4.4). The strain γ was kept constant at $5 \cdot 10^{-3}$. As described in Chapter 2.3.1 by Eq. (2.35), the relaxation modulus $G(t)$ is defined as the ratio between the resulting stress $\sigma(t)$ and the applied strain γ and can be represented by multiple Maxwell elements linked in parallel. Thus Eq. (2.48) was fitted to $G(t)$ and n was varied from 1 to 4

$$G(t) = \sum_{i}^{n=1...4} G_i e^{-t/\tau_i}. \tag{4.5}$$

* Data was partially obtained in cooperation with Steffi Günther, Institute for Particle Technology, TU Braunschweig.

The regression quality for the relaxation experiments improved with an increase in Maxwell elements from $R^2 = 0.65$ (n = 1), $R^2 = 0.92$ (n = 2), $R^2 = 0.96$ (n = 3) to $R^2 = 0.97$ (n = 4). Two elements in parallel already showed a good model to experimental data correlation. When fitted with the same parameters to the frequency sweep data with three and four Maxwell elements in parallel led to a acceptable agreement with the storage modulus G', however, only a four element Maxwell-Wiechert-model resulted in an a good accordance of the experimental data with the computed values for the loss modulus G''. The obtained parameters for the spring constants G_i and relaxation times τ_i are shown in Table 4.5. As discussed in Chapter 4.3.8, the relaxation time constants τ_i and the spring constants G_i of the biofilm measurement were in a similar range. A corresponding improvement of the prediction with respect to G' and G'' was observed with an increase in the number of parallel Maxwell elements and led to a significant consistency with the worm-like-chain-model proposed by Ehret et al. [41], who assumed that four types of attraction, i.e., electrostatic, dispersion forces, hydrogen bonds and entanglements, give the EPS its stability.

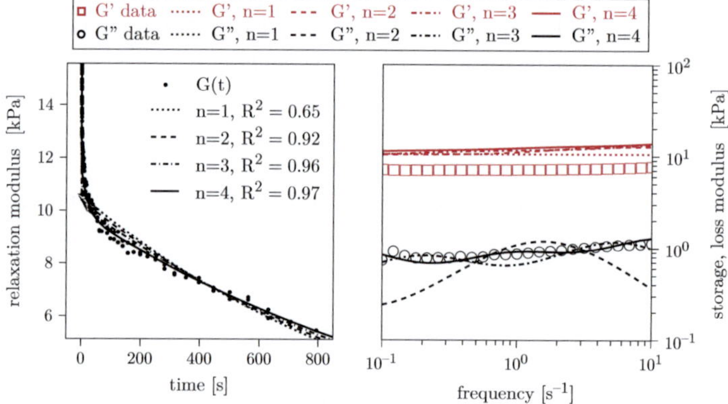

Figure 4.28: Relaxation (left) and frequency sweep (right) experiments for the hydrogel #15 (CP). Applied relaxation strain and oscillatory strain were constant with $\gamma = 5 \cdot 10^{-4}$ and $\gamma = \pm 5 \cdot 10^{-4}$, respectively. The relaxation modulus was fitted to Eq. (2.48) and the storage and loss moduli were fitted to Eqs. (2.49) and (2.50) with n = 1...4.

Graessley et al. [55], Kong et al. [78] and Stevens et al. [147] declared similar attractions being responsible for the hydrogel structure and Tang et al. [159] obtained comparable results in their stress relaxation experiments for gellan. Hence, these results were used for a superposition of the worm-like-chains and resulted in a network model for hydrogels based on gellan. The values received for the spring constants and relaxation times constants were within the same dimensions as those of the biofilms for the first three Maxwell elements (compare Chapter 4.3.8). An increase of one magnitude was observed in the relaxation time for the fourth Maxwell element for *E. coli* K12 but not for *P. putida* KT2440 biofilms, which probably was a result of longer polymer chains in the EPS of *E. coli* K12 biofilms compared to the hydrogel-matrix or *P. putida* KT2440 biofilms.

Table 4.5: Parameters of Maxwell-Wiechert-model with four Maxwell elements for a gellan based hydrogel #15 (CP).

Maxwell element i	G_i [kPa]	τ_i [s]
1	2.400	0.076
2	1.316	1.064
3	1.703	23.35
4	9.864	1334

4.4.7 Influence of Sodium and Magnesium Ions on the Sol-Gel Temperature of Gellan Gum

For the application of the physico-chemical biofilm model on the basis of thermo-reversible gellan based hydrogel it is important to know the sol-gel temperature (T_{SG}). Thus, G' and G'' were measured for each hydrogel during cooling. The liquid gellan solutions were cooled down until they solidified. This resulted in a sudden increase in G' and G''. The sol-gel temperature was defined as the inflection point of the G' and G'' curves. These were influenced by the amount of x_{Gellan}, $c_{Mg^{2+}}$ and c_{Na^+} but not by the frequency. Thus, the same approach as in Chapter 4.4.4 was used. Eq. (3.4) was adopted and the frequency as factor removed. The response R was given as the sol-gel temperature T_{SG}

$$T_{SG} = a_0 + a_1 x_{Gellan} + a_2 c_{Mg^{2+}} + a_3 c_{Na^+}$$
$$+ a_4 x_{Gellan} c_{Mg^{2+}} + a_5 x_{Gellan} c_{Na^+} + a_6 c_{Mg^{2+}} c_{Na^+} \qquad (4.6)$$
$$+ a_7 x_{Gellan}^2 + a_8 c_{Mg^{2+}}^2 + a_9 c_{Na^+}^2.$$

The presented response surfaces are the result of fitting T_{SG} of the 15 hydrogels of different compositions to Eq. (4.6). This resulted in a suggested model with $F = 53.28$ and an adequate precision ratio of 23.59. The significant model terms were x_{Gellan}, $c_{Mg^{2+}}$, c_{Na^+}, $x_{Gellan}c_{Mg^{2+}}$, $x_{Gellan}c_{Na^+}$, x_{Gellan}^2, and $c_{Na^+}^2$, respectively. The empirical relation is given by

$$\sqrt{T_{SG}} = 50.828 - 18.648 x_{Gellan} - 2.964 c_{Mg^{2+}} - 0.097 c_{Na^+}$$
$$- 0.860 x_{Gellan} c_{Mg^{2+}} + 0.172 x_{Gellan} c_{Na^+} \qquad (4.7)$$
$$+ 13.607 x_{Gellan}^2 + 0.281 c_{Mg^{2+}}^2.$$

Eq. (4.7) allows to predict T_{SG} for any gellan based hydrogel within the design space or compute the gellan content x_{Gellan}, the ion concentrations $c_{Mg^{2+}}$ and c_{Na^+} at a desired T_{SG}. The influences of the gellan and the ions on T_{SG} were quadratic and are shown in Figs. 4.29 and 4.30. T_{SG} had a clear minimum at x_{Gellan} around 0.7 %(w/v) and $c_{Mg^{2+}}$ between 6 and 7 mM but hardly changed by an alteration of the amount of c_{Na^+}. This contrasts with Miyoshi et al. [101, 102], who described a direct connection of the T_{SG} to the salt concentration.

Izumi et al. [68] published data which showed that the thermal transition curve was insensitive to the type of salt while the mechanical transition was described as very sensitive. The observed sol-gel transition correlates with the mechanical transition, thus be insensitive to the salt which would agree with the influence of c_{Na^+}. Since $c_{Mg^{2+}}$ is a divalent ion, its influence is larger at smaller concentrations. This might explain the differences in the observation between c_{Na^+} and $c_{Mg^{2+}}$. At the small concentration changes of c_{Na^+} no change in T_{SG} was observed, but the modifications in x_{Gellan} and $c_{Mg^{2+}}$ were large enough in these experiments to shift T_{SG} in either direction.

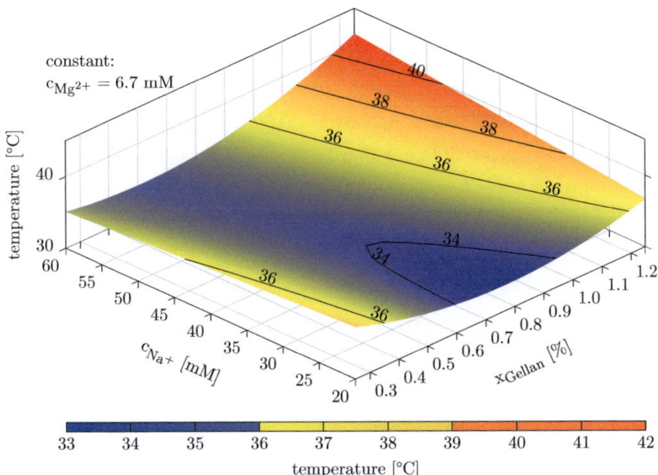

Figure 4.29: Results of the response surface methodology for the sol-gel temperature (T_{SG}). c_{Na^+} and x_{Gellan} were varied, $c_{Mg^{2+}} = 6.7$ mM was kept constant.

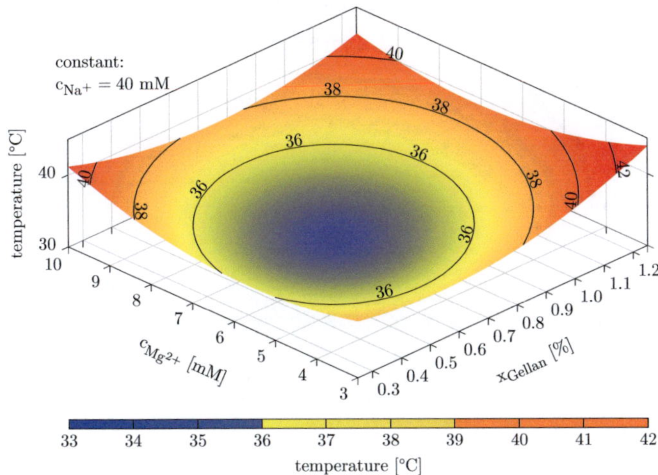

Figure 4.30: Results of the response surface methodology for the sol-gel temperature (T_{SG}). $c_{Mg^{2+}}$ and x_{Gellan} were varied, $c_{Na^+} = 40$ mM was kept constant.

4.4.8 Oxygen Diffusion in a Biomimetic Hydrogel*

To simulate biofilms with hydrogels and later label them with dead or living cells the hydrogels have to allow a similar oxygen supply. The diffusion coefficient (D_F) in the hydrogel #15 (CP) was estimated at 30 °C by measuring the concentration of the dissolved oxygen at different depths in the hydrogel with a microelectrode (compare Chapter 3.2.5). At a constant flow rate, the hydrogel was stripped of all oxygen by sparging the water with nitrogen. Then nitrogen was turned off and air was bubbled into the water to saturate it with oxygen. A fast flow rate and oxygen saturated water allowed to assume that convective mass transport was not limiting. Thus, after passing the boundary layer the oxygen concentration within a hydrogel depended on diffusion, only [142]. Fig. 4.31 shows the dissolved oxygen profiles at a certain hydrogel depth.

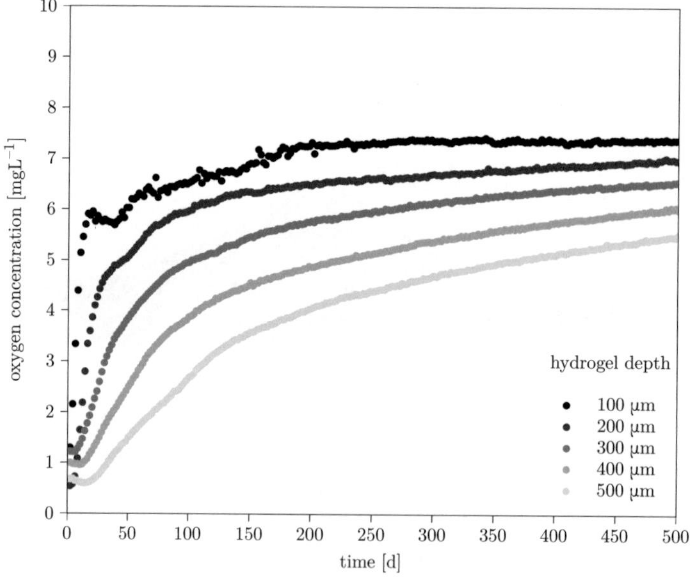

Figure 4.31: Oxygen profiles in hydrogel # 15.

---

* Data was partially obtained in cooperation with Tim Nadrowski (2013) as part of his bachelor thesis, TU Braunschweig.

Fick's second law (Eq. (2.30)) was set up for each data point i as described in Chapter 2.2.4 and solved via the least square method for the diffusion coefficient (D_F) and the oxygen concentration (c^*).

$$0 = \sum_i^n c(x_i, t_i) - c^* - (c^* - c_0)\mathrm{err}\left(\frac{x}{2\sqrt{D_F t_i}}\right). \tag{4.8}$$

The diffusion coefficient was computed to be $D_F = 1.60 \cdot 10^{-9}$ m^{2}s^{-1} with an estimated oxygen bulk concentration of $c^* = 7.97$ mgL^{-1}, being only 5.8 % higher than the tabled saturation concentration of 7.53 mgL^{-1} at 30 °C. Hence, it was assumed that c^* was equal to the saturation concentration of the bulk water, $c^* = 7.53$ mgL^{-1}, resulting in an increased diffusion coefficient of $D_F = 2.02 \cdot 10^{-9}$ m^{2}s^{-1}. This resulted to an oxygen diffusion coefficient within the range of $D_F = 1.60 \cdot 10^{-9}$ to $2.02 \cdot 10^{-9}$ m^{2}s^{-1} being slightly lower than in water with $D_F = 2.29 \cdot 10^{-9}$ m^{2}s^{-1} [50]. The results were comparable to data published by Fu et al. [50] for biofilms and agreed with data stated by Hyust et al. [64], who investigated oxygen diffusion in calcium alginate, $\varkappa$-carrageenan, gellan, agar and agarose hydrogels and reported a reduction in oxygen diffusion with an increased polymer concentration. Values for the D_F reported in literature reached from $D_F = 1.5 \cdot 10^{-9}$ to $D_F = 2.1 \cdot 10^{-9}$ m^{2}s^{-1} for hydrogels with 0.5 to 5 %(w/v) polymer [64] or $D_F = 2.3 \cdot 10^{-9}$ m^{2}s^{-1} for biofilms [73].

4.5 Proof of Principle*

The viscoelastic behavior of the biofilm was determined by oscillation-frequency and relaxation tests analogous to the gellan hydrogels. Figs. 4.16 to 4.18 and 4.28 show that identical model approaches can be used to describe the viscoelastic behavior of the biofilms and hydrogels, thus as postulated it is possible to simulate biofilm behavior with gellan based hydrogels. A physico-chemical model on gellan based hydrogel was developed with a surface response methodology. The obtained Eqs. (4.3) and (4.4) allowed to compute the concentrations of x_{Gellan}, $c_{Mg^{2+}}$ and c_{Na^+} of a hydrogel based on data for G' and G'' to mimic biofilm mechanics in terms of G' and G''.

Therefore, a *P. putida* based biofilm was characterized mechanically in terms of G' and G'' which were taken to compute the model hydrogel. Fig. 4.32 shows the

---

* Data was partially obtained in cooperation with Steffi Günther, Institute for Particle Technology, TU Braunschweig.

dependency of G' and G'' on the frequency at a strain of $5 \cdot 10^{-3}$, approximately within the linear viscoelastic range for a *P. putida* KT2440 sample. The variations in the standard error are noticeable, which can be explained by the higher variability in the biofilm density and structure. The viscoelastic behavior of biofilms corresponded to the gellan hydrogels and showed also an almost Hookean solid-like behavior. The frequency sweep yielded a G' of 11.96 ± 2.41 kPa and a G'' of 0.94 ± 0.26 kPa at 1 s^{-1}, which was in a good agreement with the hydrogels.

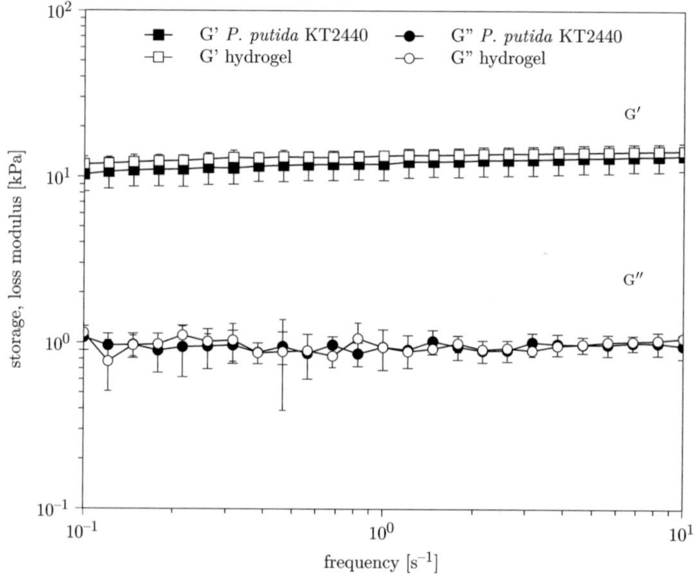

Figure 4.32: G' and G'' versus frequency. The values for the *P. putida* KT2440 biofilms (solid) were averaged over six runs and the values of the hydrogels (white) were measured as triplicates. The G' and G'' of the biofilms were used to compute the equivalent composition of a hydrogel with the help of the empirical model resulting in the following composition: $x_{\text{Gellan}} = 0.75$ %(w/v), $c_{\text{Mg2+}} = 9.4$ mM and $c_{\text{Na+}} = 26$ mM.

Furthermore, the obtained data for the *P. putida* KT2440 biofilm suggested that this biofilm could be simulated with the concept provided. Its viscoelastic parameters, G' and G'', were within the range of the measured values for the CCD. Data is shown in Table 4.6.

To investigate the prediction quality of the empirical physico-chemical model, the compositions of three random hydrogels with arbitrarily chosen values for G' and G'' were computed and rheologically characterized in terms of G' and G'' (Table 4.6). Comparing the measured and the computed data, an error of 6.5 to 12.7 % for G' was obtained. This was within the error of the rheometer that was given with up ± 10 % [96]. The error for the loss modulus was in a range from 0.2 to 17.0 %, thus higher but for a biological system still reasonable. This ensured, that the provided model is adequate to predict the general behavior of the gellan hydrogels within the design space.

Table 4.6: Predicted versus measured values for G' and G'', shown for three random hydrogels and a predicted biomimetic hydrogel. x_{Gellan}, $c_{Mg^{2+}}$ and c_{Na^+} at a constant frequency of $1\ s^{-1}$ are computed based on the provided empirical model for the shown G' and G'' values. Viscoelastic data of a *P. putida* KT2440 biofilm grown on membrane filter is given for comparison.

Gel composition			G' [kPa]	G'' [kPa]
x_{Gellan} [%(w/v)]	$c_{Mg^{2+}}$ [mM]	c_{Na^+} [mM]	computed/measured/error	
0.86	9.5	23	16.05 / 14.68 / 8.5 %	1.27 / 1.33 / 4.2 %
0.97	6.4	50	7.06 / 7.63 / 8.1 %	0.53 / 0.61 / 17.0 %
0.88	5.0	50	4.74 / 5.04 / 6.5 %	0.33 / 0.28 / 15.9 %
Model			G' [kPa]	G'' [kPa]
0.75	9.4	26	11.96 / 13.41 / 12.7 %	0.94 / 0.94 / 0.2 %
Biofilm			G' [kPa]	G'' [kPa]
P. putida KT2440			11.96 ± 1.39	0.94 ± 0.15

4.5.1 Worm-Like-Chain-Model for Gellan

As already shown in Chapters 4.3.8 and 4.4.6 the biofilm and the hydrogel can both be modeled with the worm-like-chain-model. According to Ehret et al. [41] the relaxation times τ_i are characteristic for the type of junction and do not vary with changing $c_{Mg^{2+}}$ levels. It was assumed, that this was also valid for c_{Na^+}. Thus, Eqs. (2.49) and (2.50) were adopted and G' and G'' of all 15 hydrogels were fitted to these equations with constant relaxation times τ_i for all hydrogels but with changing spring constants $G_{k,i}$,

$$G'_k(\omega) = \sum_{i=1}^{4} G_{k,i} \frac{\omega^2 \tau_i^2}{1 + \omega^2 \tau_i^2} \tag{4.9}$$

and

$$G''_k(\omega) = \sum_{i=1}^{4} G_{k,i} \frac{\omega \tau_i}{1 + \omega^2 \tau_i^2} \ . \tag{4.10}$$

The index i represents the Maxwell elements (i = 1...4) and k identifies the # of the hydrogel as used in Table 4.4. For the estimation of τ_i and $G_{k,i}$ the sum of the squared errors between the frequency dependent G' and G'' was minimized with Matlab. Fig. 4.33 shows an overall good agreement between the resulting model graphs for the hydrogel and the measured data.

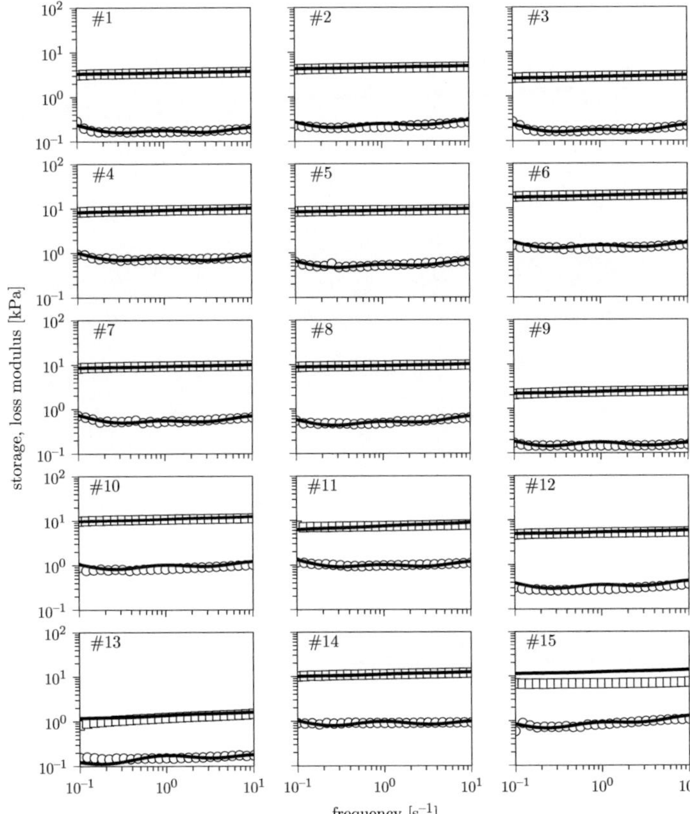

Figure 4.33: Storage (G′, square) and loss (G″, circle) moduli versus the frequency of the CCD data. Experimental mean values are presented as symbols while the lines are the with Eqs. (4.9) and (4.10) fitted values.

The results are summarized in Table A.1. Fig. 4.34 shows the color coded content of x_{Gellan}, concentrations of $c_{Mg^{2+}}$ and c_{Na^+}, the spring constants ($G_{i,\#k}$) for each Maxwell element i with i = 1...4, as well as $G'_{\#k}$ and $G''_{\#k}$ for each hydrogel #k with k = 1...15 at a frequency of 1 s^{-1}.

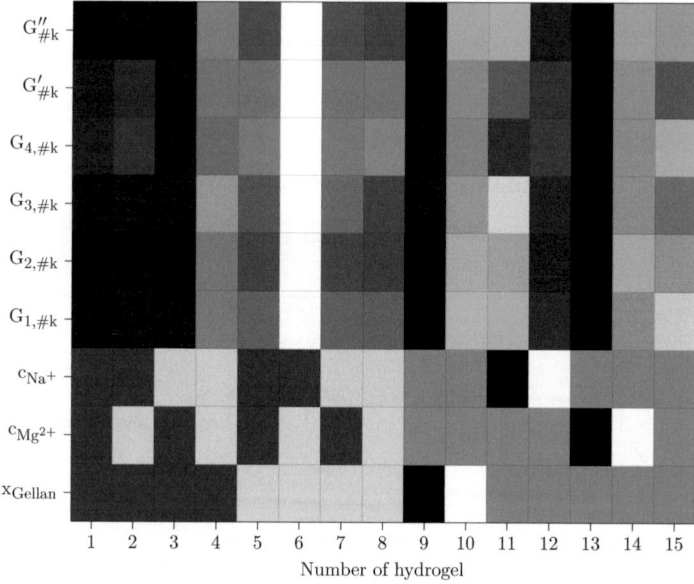

Figure 4.34: Color coded concentrations of x_{Gellan}, $c_{Mg^{2+}}$ and c_{Na^+} and spring constants ($G_{k,i}$) for each Maxwell element i = 1...4 as well as $G'_{\#k}$ and $G''_{\#k}$ for each hydrogel #k with k = 1...15 at a frequency of 1 s^{-1}. The highest value is coded black and the lowest white. The numbers correlate with hydrogels in Table 4.4

The highest spring constants were found at the hydrogels #9 and #13 which correlated with the maximum G$'$ and G$''$ values. Hydrogel #9 had the most x_{Gellan} while #13 had the highest $c_{Mg^{2+}}$. In contrast, the values for $G_{i,\#k}$, G$'$ and G$''$ for the hydrogel #11, having the maximum c_{Na^+}, were rather small. A similar tendency was observed when comparing the hydrogels #1...#4, #5...#8 and #11...#15 with identical x_{Gellan}. Low amounts of $c_{Mg^{2+}}$ resulted in small $G_{i,\#k}$, $G'_{\#k}$ and $G''_{\#k}$ values, high amounts increased the overall strength of the gel. The weakening effect of high concentrations of Na$^+$ was confirmed as well as the strengthening effect of Gellan and Mg^{2+}. All hydrogels showed a similar tendency, high G' correlated with high G'' values. These results show an overall good agreement between the measured values for G' and G'', the empirical model developed on basis of the frequency sweep with the surface response methodology and the data obtained from the physical model as well as the superposition of the worm-like-chain-model.

5 Conclusion

Research in biofilm and biofilm mechanics has, based on the number of publications, increased ever since it started [45]. However, biofilm systems and their mechanics are still not entirely understood. The results shown are a contribution and provide a tool to increase the general understanding of biofilm systems.

The main objective of this work was the development of a hydrogel based physico-chemical biofilm model to investigate fluid mechanical interaction and its influence on growth and detachment effects of biofilms. Therefore, representative and reproducible single culture biofilms were required. Additionally, established measurement techniques had to be validated with either real biofilms or with the help of the physico-chemical biofilm model based on highly hydrolyzed gellan hydrogels.

The growth experiments clearly showed that controlled biofilm development is still challenging and difficult. The established microbial test system based on the BTR allowed to monitor and characterize biofilm growth. The biofilms grew either on the walls of the tube, on surface modified object slides, on plastic and iron nettings or on membrane filters.

Automated on-line analytics were established and dissolved oxygen, pH, temperature and planktonic cell growth were monitored. Image acquisition of surface attached biofilms was introduced and adapted to visualize biofilm development in terms of growth and detachment processes. The real biofilms and the hydrogel model were used to validated the established measurement techniques. The introduced relative biofilm coverage allowed to quantitatively identify growth and detachment periods during the biofilm development. However, it only contained limited information about the biofilm thickness and required constant illumination in order to be used as comparison tool between biofilms. Experiments suggested that the development of a biofilm was partially by chance. It was not sufficient to control the environmental and culturing parameters to obtain reproducible biofilms.

For future visualization of cell distribution within biofilms and hydrogels with the confocal laser scanning microscopy the *P. putida* KT2440 strain was successfully genetically modified to produce the green fluorescent protein (GFP).

Model biofilms based on a mucoid bacterial lawn and grown on membrane filters were mechanically characterized in terms of the storage, loss and relaxation moduli.

The values measured were successfully related to the proposed worm-like-chain-model developed by Ehret et al. [41]. The resemblance of a biofilm to a hydrogel was shown by comparing gellan gum to biofilms. A Central Composite Design was applied to obtain an empirical model of the viscoelastic properties in terms of storage (G') and loss (G'') moduli depending on the gellan, Na^+ and Mg^{2+} concentrations of the hydrogel. The model was statistically verified and compared to experimental data and showed a good consistency between the empirical model and biofilms. Furthermore, it was verified that the worm-like-chain-model can be used to simulate the hydrogel and the biofilm data, demonstrating a strong analogy between the empirical and physical model and the experimental data.

This proved that the biofilm production can be simplified by using a physico-chemical model based on highly hydrolyzed gellan hydrogels. The results demonstrated that the response surface methodology is a reliable tool to predict the mechanical properties of gellan based hydrogels. The obtained empirical physico-chemical models for the storage (G') and loss (G'') moduli provide a tool that can be used to adjust and fine tune gel properties to mimic biofilm mechanical behavior. The results indicated a good estimation of the viscoelastic characteristic values of G' and G'' in the frequency interval of 0.1 to 10 s^{-1}. The G' and G'' values of the predicted and designed hydrogels differed by less than 17.0 % and were similar to those of the *P. putida* KT2440 biofilm, conclusively demonstrating that this method enables the design of gellan based hydrogels which mimic biofilm behavior. Additional results also provided a tool to estimate the sol-gel temperature of the designed hydrogel for further applications and showed similar oxygen diffusivity in comparison to biofilms.

6 References

[1] Adav, S. S., Lee, D.-J., Tay, J.-H. (2008). Extracellular polymeric substances and structural stability of aerobic granule. Water Res. 42, 6-7, 1644–1650.

[2] Allison, D. G., Sutherland, I. (1987). The role of exopolysaccharides in adhesion of freshwater bacteria. J. Gen. Microbiol. 133, 1319–1327.

[3] Alpkvist, E., Klapper, I. (2007). Description of mechanical response including detachment using a novel particle model of biofilm/flow interaction. Water Sci. Technol. 55, 2001, 265–273.

[4] Alpkvist, E., Overgaard, N. C., Gustafsson, S., Heyden, A. (2004). A new mathematical model for chemotactic bacterial colony growth. Water Sci. Technol. 49, 11-12, 187–192.

[5] Alpkvist, E., Picioreanu, C., van Loosdrecht, M. C. M., Heyden, A. (2006). Three-dimensional biofilm model with individual cells and continuum EPS matrix. Biotechnol. Bioeng. 94, 5, 961–979.

[6] Anseth, K. S., Bowman, C. N., Brannon-Peppas, L. (1996). Mechanical properties of hydrogels and their experimental determination. Biomaterials 17, 1647–1657.

[7] Atkinson, A. C. (2006). Generalized linear models and response transformation. In: Response surface methodology and related topics (Khuri, A. I. (ed.)). World Scientific, New Jersey-London-Singapore-Beijing-Shanghai-Hong Kong-Taipei-Chennai.

[8] Bajaj, I. B., Survase, S. A. (2007). Gellan gum: fermentative production, downstream processing and applications. Food Technol. 45, 4, 341–354.

[9] Bales, P. M., Renke, E. M., May, S. L., Shen, Y., Nelson, D. C. (2013). Purification and Characterization of Biofilm-Associated EPS Exopolysaccharides from ESKAPE Organisms and Other Pathogens. PloS one 8, 6, 1–8.

[10] Beucher, O. (2005). Wahrscheinlichkeitsrechnung und Statistik mit MATLAB. 2 ed. Berlin Heidelberg: Springer-Verlag.

[11] Blanch, H. W., Clark, D. S. (1997). Biochemical Engineering. 2nd ed. Marcel Dekker, New York.

[12] Böl, M., Ehret, A. E., Bolea Albero, A., Hellriegel, J., Krull, R. (2013). Recent advances in mechanical characterisation of biofilm and their significance for material modelling. Crit. Rev. Biotechnol. 33, 2, 145–171.

[13] Böl, M., Möhle, R. B., Haesner, M., Neu, T. R., Horn, H., Krull, R. (2009). 3D finite element model of biofilm detachment using real biofilm structures from CLSM data. Biotechnol. Bioeng. 103, 1, 177–186.

[14] Bolea Albero, A., Ehret, A. E., Böl, M. (2014). A new approach to the simulation of microbial biofilms by a theory of fluid-like pressure-restricted finite growth. Comput. Method. Appl. M. 272, 271–289.

[15] Braccini, I., Pérez, S. (2001). Molecular basis of C(2+)-induced gelation in alginates and pectins: the egg-box model revisited. Biomacromolecules 2, 4, 1089–1096.

[16] Bryers, J. D. (2000). Biofilms: An introduction. In: Biofilms II: process analysis and applications (Mitchell, R. (ed.)) 1st ed. John Wiley & Sons, New York.

[17] Byskov, E. (2013). Linear elasticity. In: Elementary Continuum Mechanics for Everyone: With Applications to Structural Mechanics (Gladwell, G. M. L. (ed.)). Springer, Berlin-Heidelberg.

[18] Carl Roth GmbH & Co. KG (2006). GELRITE - Gellan Gum for Microbiological Applications. Karlsruhe.

[19] Celik, G. Y., Aslim, B., Beyatli, Y. (2008). Characterization and production of the exopolysaccharide (EPS) from *Pseudomonas aeruginosa* G1 and *Pseudomonas putida* G12 strains. Carbohydr. Polym. 73, 1, 178–182.

[20] Cesàro, A., Gamini, A., Navarini, L. (1992). Supramolecular structure of microbial polysaccharides in solution: from chain conformation to rheological properties. Polymer 33, 19, 4001–4008.

[21] Cesàro, A., Sussich, L., Navarini, L. (2004). Order-disorder conformational transitions of carbohydrate polymers. In: The Nature of Biological Systems as Revealed by Thermal Methods (Lörinczy, D. (ed.)). Springer, Berlin-Heidelberg.

[22] Chambless, J., Stewart, P. (2007). A three dimensional computer model analysis of three hypothetical biofilm detachment mechanisms. Biotechnol. Bioeng. 97, 6, 1573–1584.

[23] Chandrasekaran, R., Millane, R. (1988). The crystal structure of gellan. Carbohydr. 175, 1–15.

[24] Chandrasekaran, R., Thailambal, V. (1990). The influence of calcium ions, acetate and L-glycerate groups on the gellan double-helix. Carbohydr. Polym. 12, 4, 431–442.

[25] Chang, W.-S., van de Mortel, M., Nielsen, L., Nino de Guzman, G., Li, X., Halverson, L. L. J. (2007). Alginate production by *Pseudomonas putida* creates a hydrated microenvironment and contributes to biofilm architecture and stress tolerance under water-limiting conditions. J. Bacteriol. 189, 22, 8290–8299.

[26] Characklis, W. G. (1973). Attached microbial growths - II. Frictional resistance due to microbial slimes. Water Res. 7, 9, 1249–1258.

[27] Characklis, W. G. (1981). Fouling Biofilm Development : A Process Analysis. Biotechnol. Bioeng. 102, 2, 1923–1960.

[28] Characklis, W. G., Marshall, K. C. (1990). Biofilms Vol. 5 of Wiley Series in Ecological and Applied Microbiology. Wiley.

[29] Cheng, L., Xia, X., Yu, W., Scriven, L. E., Gerberich, W. W. (2000). Flat-Punch Indentation of Viscoelastic Material. J. Polym. Sci. Phys. 38, 1, 10–22.

[30] Chippada, U., Yurke, B., Georges, P. C., Langrana, N. a. (2009). A nonintrusive method of measuring the local mechanical properties of soft hydrogels using magnetic microneedles. J. Biomech. Eng. 131, 2, 021014–1–021014–12.

[31] Chung, T.-P., Tseng, H.-Y., Juang, R.-S. (2003). Mass transfer effect and intermediate detection for phenol degradation in immobilized *Pseudomonas putida* systems. Process Biochem. 38, 10, 1497–1507.

[32] Coroneo, M., Yoshihara, L., Wall, W. A. (2014). Biofilm growth: A multi-scale and coupled fluid-structure interaction and mass transport approach. Biotechnol. Bioeng 9999, 1–11.

[33] Corr, D. T., Starr, M. J., Vanderby, R., Best, T. M. (2001). A Nonlinear Generalized Maxwell Fluid Model for Viscoelastic Materials. J. Appl. Mech. 68, 5, 787–790.

[34] Costerton, J. W. (1999). Introduction to biofilm. Int. J. Antimicrob. Agents 11, 3-4, 217–239.

[35] Coutinho, D. F., Sant, S. V., Shin, H., Oliveira, J. T., Gomes, M. E., Neves, N. M., Khademhosseini, A., Reis, R. L. (2010). Modified Gellan Gum hydrogels with tunable physical and mechanical properties. Biomaterials 31, 29, 7494–7502.

[36] Del Nobile, M. A., Chillo, S., Mentana, A., Baiano, A., Nobile, M. A. D. (2007). Use of the generalized Maxwell model for describing the stress relaxation behavior of solid-like foods. J. Food Eng. 78, 3, 978–983.

[37] Di Stefano, A., D'Aurizio, E., Trubiani, O., Grande, R., Di Campli, E., Di Giulio, M., Di Bartolomeo, S., Sozio, P., Iannitelli, A., Nostro, A., Cellini, L. (2009). Viscoelastic properties of *Staphylococcus aureus* and *Staphylococcus epidermidis* mono-microbial biofilms. Microb. Biotechnol. 2, 6, 634–641.

[38] Dockery, J. (2001). Finger formation in biofilm layers. SIAM J. Appl. Math. 62, 3, 853–869.

[39] Doyle, J., Giannouli, P., K., P., R., M. E. (2002). Effect of K^+ and Ca^{2+} cations on gellation of ϰ-carrageenan. In: Gums and Stabilisers for the Food Industry (Williams, P. A. , Phillips, G. (eds.)) Vol. 11. The Royal Society of Chemistry, Cambridge.

[40] Duffrène, L., Gy, R., Burlet, H., Piques, R., Faivre, A., Sekkat, A., Perez, J. (1997). Generalized Maxwell model for the viscoelastic behavior of a soda-lime-silica glass under low frequency shear loading. Rheol. Acta 36, 173–186.

[41] Ehret, A. E., Böl, M. (2012). Modelling mechanical characteristics of microbial biofilms by network theory. J. R. Soc. Interface 1–27.

[42] EIFAC (1986). Technical paper. European Inland Fisheries Advisory Commission, Food and Agriculture Organization of the United Nations.

[43] El-Bassi, L., Iwasaki, H., Oku, H., Shinzato, N., Matsui, T. (2010). Biotransformation of benzothiazole derivatives by the *Pseudomonas putida* strain HKT554. Chemosphere 81, 1, 109–113.

[44] Elasri, M. O., Miller, R. V. (1999). Study of the Response of a Biofilm Bacterial Community to UV Radiation. Appl. Environ. Microbiol. 65, 5, 2025–2031.

[45] Elsevier B.V. (2014). Scopus. Retrieved 3/4/2014, from http://www.scopus.com.

[46] Ferry, J. D. (1980). Viscoelastic properties of polymers. 3rd ed. John Wiley & Sons, New York.

[47] Flemming, H.-C., Schaule, G. (1994). Mikrobielle Werkstoffzerstörung - Biofilm und Biofouling: Bekämpfung von Biofouling in wäßrigen Systemen. Mater. Corros. und Korrosion 45, 1, 40–53.

[48] Flemming, H.-C., Wingender, J. (2003). Extracellular Polymeric Substances (EPS): Structural, Ecological and Technical Aspects. John Wiley & Sons, New York.

[49] Frenzel, F., B. Gebhard (2009). Stofftransport. In: Physik Formelsammlung: mit Erläuterungen und Beispielen aus der Praxis für Ingenieure und Naturwissenschaftler (Kurzwei, P. (ed.)). Springer, Berlin-Heidelberg.

[50] Fu, Y., Zhang, T., Bishop, P. (1994). Determination of effective oxygen diffusivity in biofilms grown in a completely mixed biodrum reactor. Water Sci. Technol. 29, 10-11, 455–462.

[51] García, M. C., Alfaro, M. C., Calero, N., Muñoz, J. (2011). Influence of gellan gum concentration on the dynamic viscoelasticity and transient flow of fluid gels. Biochem. Eng. J. 55, 2, 73–81.

[52] Giavasis, I., Harvey, L. M., McNeil, B. (2000). Gellan gum. Crit. Rev. Biotechnol. 20, 3, 177–211.

[53] Gibas, I., Janik, H. (2010). Synthetic polymer hydrogels for biomedical applications. Chem. Chem. Technol. 4, 4, 297–304.

[54] Gilmour, S. G., Trinca, L. A. (2006). Response surface experiments on processes with high variation. In: Response surface methodology and related topics (Khuri, A. I. (ed.)). World Scientific, New Jersey-London-Singapore-Beijing-Shanghai-Hong Kong-Taipei-Chennai.

[55] Graessley, W. W. (1974). Molecular entanglement theories of linear viscoelastic behavoir. In: The Entanglement Concept in Polymer Rheology. Springer, Berlin-Heidelberg.

[56] Grant, G., Morris, E., Rees, D., Smith, P., Thom, D. (1973). Biological interactions between polysaccharides and divalent cations: the egg-box model. FEBS Lett. 32, 1, 195–198.

[57] Gross, R., Buehler, K., Schmid, A. (2013). Engineered catalytic biofilms for continuous large scale production of n-octanol and (S)-styrene oxide. Biotechnol. Bioeng. 110, 2, 424–436.

[58] Gross, R., Lang, K., Bühler, K., Schmid, A. (2010). Characterization of a biofilm membrane reactor and its prospects for fine chemical synthesis. Biotechnol. Bioeng. 105, 4, 705–717.

[59] Hanahan, D. (1983). Studies on transformation of *Escherichia coli* with plasmids. J. Mol. Biol. 166, 4, 557–580.

[60] Hohne, D. N., Younger, J. G., Solomon, M. J. (2009). Flexible microfluidic device for mechanical property characterization of soft viscoelastic solids such as bacterial biofilms. Langmuir 25, 13, 7743–7751.

[61] Horn, H. (1994). Dynamics of a nitrifying bacteria population in a biofilm controlled by an oxygen microelectrode. Water Sci. Technol. 29, 10-11.

[62] Howell, J. A., Atkinson, B. (1976). Sloughing of microbial film in trickling filters. Water Res. 10, 4, 307–315.

[63] Hudson, D., Margaritis, A. (2013). Biopolymer nanoparticle production for controlled release of biopharmaceuticals. Crit. Rev. Biotechnol. , October, 1–19.

[64] Hulst, A. C., Hens, H. J. H., Buitelaar, R. M., Tramper, J. (1989). Determination of the effective diffusion coefficient of oxygen in gel materials in relation to gel concentration. Biotechnol. Tech. 3, 3, 199–204.

[65] Hunt, S. M. (2003). A computer investigation of chemically mediated detachment in bacterial biofilms. Microbiology 149, 5, 1155–1163.

[66] Imeson, A. (2010). Food stabilisers, thickeners and gelling agents. Wiley-Blackwell, Hoboken.

[67] Irgens, F. (2008). Theory of elasticity. In: Continuum Mechanics. Springer, Berlin-Heidelberg.

[68] Izumi, Y., Kikuta, N., Sakai, K., Takezawa, H. (1996). Phase diagrams and molecular structures of sodium-salt-type gellan gum. Carbohydr. Polym. 3617, 96, 121–127.

[69] Jahn, A., Griebe, T., Nielsen, P. H. (1999). Composition of *Pseudomonas putida* biofilms: Accumulation of protein in the biofilm matrix. Biofouling 14, 1, 49–57.

[70] Jansson, P.-E., Lindberg, B., Sandford, P. A. (1983). Structural studies of gellan gum, an extracellular polysaccharide elaborated by *Pseudomonas elodea*. Carbohydr. Res. 124, 1, 135–139.

[71] Jhon, M. S., Andrade, D. (1973). Water and Hydrogels. J. Biomed. Mater. Res. 7, 6, 509–522.

[72] Kachlany, S. C., Levery, S. B., Kim, J. S., Reuhs, B. L., Lion, L. W., Ghiorse, W. C., Leung, S. (2001). Structure and carbohydrate analysis of the exopolysaccharide capsule of *Pseudomonas putida* G7. Environ. Microbiol. 3, 12, 774–784.

[73] Kapellos, G. E., Alexiou, T. S., Pavlou, S., Payatakes, A. C. (2004). Hierarchical modelling approach for the prediction of effective hydraulic permeability and diffusion coefficient in biofilms. Intern. Conf. Biofilms 255–260.

[74] Kawahara, S., Yoshikawa, A., Hiraoki, T., Tsutsumi, A. (1996). Interactions of paramagnetic metal ions with gellan gum studied by ESR and NMR methods. Carbohydr. Polym. 30, 2-3, 129–133.

[75] Klapper, I., Dockery, J. (2006). Role of cohesion in the material description of biofilms. Phys. Rev. E 74, 3, 1–8.

[76] Klapper, I., Rupp, C. J., Cargo, R., Purevdorj, B., Stoodley, P. (2002). Viscoelastic fluid description of bacterial biofilm material properties. Biotechnol. Bioeng. 80, 3, 289–296.

[77] Klausen, M., Gjermansen, M., Kreft, J. U., Tolker-Nielsen, T. (2006). Dynamics of development and dispersal in sessile microbial communities: examples from *Pseudomonas aeruginosa* and *Pseudomonas putida* model biofilms. FEMS Microbiol. Lett. 261, 1, 1–11.

[78] Kong, H. (2002). Decoupling the dependence of rheological/mechanical properties of hydrogels from solids concentration. Polymer 43, 23, 6239–6246.

[79] Körstgens, V. (2003). Die mechanische Stabilität bakterieller Biofilme: Eine Untersuchung verschiedener Einflüsse auf Biofilme von *Pseudomonas aeruginosa*. PhD thesis, Universität Duisburg - Essen.

[80] Körstgens, V., Flemming, H.-C., Wingender, J., Borchard, W. (2001). Influence of calcium ions on the mechanical properties of a model biofilm of mucoid *Pseudomonas aeruginosa*. Water Sci. Technol. 43, 6, 49–57.

[81] Körstgens, V., Flemming, H.-C., Wingender, J., Borchard, W. (2001). Uniaxial compression measurement device for investigation of the mechanical stability of biofilms. J. Microbiol. Methods 46, 1, 9–17.

[82] Kreyenschulte, D., Krull, R., Margaritis, A. (2014). Recent advances in microbial biopolymer production and purification. Crit Rev Biotechnol 34, 1, 1–15.

[83] Kumar, A. S., Mody, K., Jha, B. (2007). Bacterial exopolysaccharides – a perception. J. Basic Microbiol. 47, 2, 103–117.

[84] Kuo, C. K., Ma, P. X. (2001). Ionically crosslinked alginate hydrogels as scaffolds for tissue engineering: part 1. Structure, gelation rate and mechanical properties. Biomaterials 22, 6, 511–521.

[85] Laaman, T. R. (2011). Hydrocolloids: Fifteen practical tips. In: Hydrocolloids in Food Processing Hydrocolloids (Laaman, T. R. (ed.)). Wiley-Blackwell, Hoboken.

[86] Lakes, R. (2009). Viscoelastic Materials. Cambridge University Press, Cambridge.

[87] Lembre, P., Lorentz, C., Martino, P. D. (2012). Exopolysaccharides of the Biofilm Matrix: A Complex Biophysical World. In: The Complex World of Polysaccharides (Karunaratne, D. N. (ed.)). InTech.

[88] Lieleg, O., Caldara, M., Baumgärtel, R., Ribbeck, K. (2011). Mechanical robustness of *Pseudomonas aeruginosa* biofilms. Soft Matter 7, 7, 3307–3314.

[89] Lin, D. K., Peterson, J. J. (2006). Statistical inference for response surface optima. In: Response surface methodology and related topics (Khuri, A. I. (ed.)). World Scientific, New Jersey-London-Singapore-Beijing-Shanghai-Hong Kong-Taipei-Chennai.

[90] Liu, K., VanLandingham, M. R., Ovaert, T. C. (2009). Mechanical characterization of soft viscoelastic gels via indentation and optimization-based inverse finite element analysis. J. Mech. Behav. Biomed. Mater. 2, 4, 355–363.

[91] Liu, L. (1995). Induction of Systemic Resistance in Cucumber Against Fusarium Wilt by Plant Growth-Promoting *Rhizobacteria*. Phytopathology 85, 6, 695–698.

[92] Liu, L., Wang, B., Gao, Y., Bai, T.-C. (2013). Chitosan fibers enhanced gellan gum hydrogels with superior mechanical properties and water-holding capacity. Carbohydr. Polym. 97, 1, 152–158.

[93] Machery-Nagel (2013). Plasmid DNA purification Excerpt from user manual.

[94] Macosko, C. W. (1994). Linear viscoelasticity. In: Rheology: principles, measurements, and applications. Wiley, New York-Chichester-Weinheim-Brisbane-Singapore-Toronto.

[95] Madigan, M., Clark, D., Stahl, D., Martinko, J. (2010). Pathways of discovery in microbiology. In: Brock Biology of Microorganisms 13th ed. Benjamin Cummings, San Francisco.

[96] Malvern (2013). personal notification 2013/11/29.

[97] Mann, E. E., Wozniak, D. J. (2012). *Pseudomonas* biofilm matrix composition and niche biology. FEMS Microbiol. Rev. 36, 4, 893–916.

[98] Mayer, C., Moritz, R., Kirschner, C., Borchard, W., Maibaum, R., Wingender, J., Flemming, H.-C. (1999). The role of intermolecular interactions: studies on model systems for bacterial biofilms. Int. J. Biol. Macromol. 26, 1, 3–16.

[99] Milas, M., Rinaudo, M. (1996). The gellan sol-gel transition. Carbohydr. Polym. 30, 2-3, 177–184.

[100] Miyoshi, E., Nishinari, K. (1999). Rheological and thermal properties near the sol-gel transition of gellan gum aqueous solutions. Colloid Polym. Sci. 114, 68–82.

[101] Miyoshi, E., Takaya, T., Nishinari, K. (1995). Gel-sol transition in gellan aqueous solutions. Macromol. Symp. 99, 83–91.

[102] Miyoshi, E., Takaya, T., Nishinari, K. (1996). Rheological and thermal studies of gel-sol transition in gellan gum aqueous solutions. Carbohydr. Polym. 30, 2-3, 109–119.

[103] Möhle, R. B. (2008). An analytic-synthetic approach combining mathematical modeling and experiments – towards an understanding of biofilm systems. ibvt-Schriftenreihe (Hempel, D. C. (ed.)) Vol. 35. PhD thesis, Technische Universität Braunschweig, FIT-Verlag, Paderborn.

[104] Möhle, R. B., Langemann, T., Haesner, M., Augustin, W., Scholl, Neu, T. R., Hempel, D. C., Horn, H. (2007). Structure and shear strength of microbial biofilms as determined with confocal laser scanning microscopy and fluid dynamic gauging using a novel rotating disc biofilm reactor. Biotechnol. Bioeng. 98, 4, 747–755.

[105] Moresi, M., Bruno, M., Parente, E. (2004). Viscoelastic properties of microbial alginate gels by oscillatory dynamic tests. J. Food Eng. 64, 2, 179–186.

[106] Moritaka, H., Fukuba, H., Kumeno, K., Nakahama, N., Nishinari, K. (1991). Effect of monovalent and divalent cations on the rheological properties of gellan gels. Food Hydrocoll. 4, 6, 495–507.

[107] Morris, E., Moritz, R. (1996). Conformational and rheological transitions of welan, rhamsan and acylated gellan. Carbohydr. Polym. 30, 2-3, 165–175.

[108] Mulisch, M., Welsch, U. (2010). Romeis Mikroskopische Technik. 17th ed. Spektrum Akademischer Verlag, Heidelberg.

[109] Nancharaiah, Y. V., Venugopalan, V. P., Wuertz, S., Wilderer, P. A., Hausner, M. (2005). Compatibility of the green fluorescent protein and a general nucleic acid stain for quantitative description of a *Pseudomonas putida* biofilm. J. Microbiol. Methods 60, 2, 179–187.

[110] Nancharaiah, Y. V., Wattiau, P., Wuertz, S., Bathe, S., Mohan, S. V., Wilderer, P. A., Hausner, M. (2003). Dual labeling of *Pseudomonas putida* with fluorescent proteins for in situ monitoring of conjugal transfer of the TOL plasmid. Appl. Environ. Microbiol. 69, 8, 4846–4852.

[111] Nedra, D., Castro, S. P. M., Paulín, E. G. L., Kwiatkowski, S., Edgar, S., Reyes, R. E., González, C. R., Jiménez, R. C., Herrera, O., Andrade, A. A., Ivashina, T. V., Ksenzenko, V. N., Shauna, L., Trofimova, N. N., Medvedeva, E. N., Karunaratne, D. N. (2012). The Complex World of Polysaccharides. InTech.

[112] Nelson, K. E., Weinel, C., Paulsen, I. T., Dodson, R. J., Hilbert, H., Martins dos Santos, V. a. P., Fouts, D. E., Gill, S. R., Pop, M., Holmes, M., Brinkac, L., Beanan, M., DeBoy, R. T., Daugherty, S., Kolonay, J., Madupu, R., Nelson, W., White, O., Peterson, J., Khouri, H., Hance, I., Chris Lee, P., Holtzapple, E., Scanlan, D., Tran, K., Moazzez, A., Utterback, T., Rizzo, M., Lee, K., Kosack, D., Moestl, D., Wedler, H., Lauber, J., Stjepandic, D., Hoheisel, J., Straetz, M., Heim, S., Kiewitz, C., Eisen, J. a., Timmis, K. N., Düsterhöft, A., Tümmler, B., Fraser, C. M. (2002). Complete genome sequence and comparative analysis of the metabolically versatile *Pseudomonas putida* KT2440. Environ. Microbiol. 4, 12, 799–808.

[113] Nickerson, M., Paulson, A. (2004). Rheological properties of gellan, $\varkappa$-carrageenan and alginate polysaccharides: effect of potassium and calcium ions on macrostructure assemblages. Carbohydr. Polym. 58, 1, 15–24.

[114] Nieto, M. B., Akins, M. (2011). Hydrocolloids in bakery fillings. In: Hydrocolloids in Food Processing Hydrocolloids (Laaman, T. R. (ed.)). Wiley-Blackwell, Hoboken.

[115] Nikolaev, Y. A., Plakunov, V. K. (2007). Biofilm - "City of microbes" or an analogue of multicellular organisms? Microbiol. 76, 2, 125–138.

[116] Nilsson, M., Chiang, W.-C., Fazli, M., Gjermansen, M., Givskov, M., Tolker-Nielsen, T. (2011). Influence of putative exopolysaccharide genes on *Pseudomonas putida* KT2440 biofilm stability. Environ. Microbiol. 13, 5, 1357–1369.

[117] Nishinari, K. (1996). Rheological and DSC studies on the interaction between gellan gum and konjac glucomannan. Carbohydr. Polym. 30, 2-3, 193–207.

[118] Nishinari, K. (1999). Introduction. In: Physical Chemistry and Industrial Application of Gellan Gum (Nishinari, K. (ed.)). Springer, Berlin-Heidelberg.

[119] Nishinari, K. (Ed.) (1999). Physical Chemistry and Industrial Application of Gellan Gum. Springer, Berlin-Heidelberg.

[120] Nishinari, K. (2009). Some thoughts on the definition of a gel. In: Gels: Structures, Properties, and Functions (Tokita, M. , Katsuyoshi, N. (eds.)). Springer, Berlin-Heidelberg.

[121] Nitta, Y., Takahashi, R., Nishinari, K. (2010). Viscoelasticity and phase separation of aqueous Na-type gellan solution. Biomacromolecules 11, 1, 187–191.

[122] Noda, S., Funami, T., Nakauma, M., Asai, I., Takahashi, R., Al-Assaf, S., Ikeda, S., Nishinari, K., Phillips, G. O. (2008). Molecular structures of gellan gum imaged with atomic force microscopy in relation to the rheological behavior in aqueous systems. 1. Gellan gum with various acyl contents in the presence and absence of potassium. Food Hydrocoll. 22, 6, 1148–1159.

[123] Ogawa, E. (1993). Osmotic pressure measurements for gellan gum aqueous solutions. Food Hydrocoll. 7, 5, 397–405.

[124] Ogawa, E. (1996). Influence of storage on the molecular weight of tetramethylammonium-type gellan gums. Carbohydr. Polym. 30, 145–148.

[125] Ohtsuka, A., Watanabe, T. (1996). The network structure of gellan gum hydrogels based on the structural parameters by the analysis of the restricted diffusion of water. Carbohydr. Polym. 30, 2-3, 135–140.

[126] Okamoto, T., Kubota, K. (1996). Sol-gel transition of polysaccharide gellan gum. Carbohydr. Polym. 30, 2-3, 149–153.

[127] Oliveira, J. T., Martins, L., Picciochi, R. (2010). Gellan gum: a new biomaterial for cartilage tissue engineering applications. J. Biomed. Mater. Res. A. 93, 3, 852–863.

[128] O'Neill, M. A., Selvendran, R. R., Morris, V. J. (1983). Structure of the acidic extracellular gelling polysaccharide produced by *Pseudomonas elodea*. Carbohydr. Res. 124, 1, 123–133.

[129] Pérez-Campos, S. J., Chavarría-Hernández, N., Tecante, A., Ramírez-Gilly, M., Rodríguez-Hernández, A. I. (2012). Gelation and microstructure of dilute gellan solutions with calcium ions. Food Hydrocoll. 28, 2, 291–300.

[130] Picioreanu, C., Kreft, J. U., van Loosdrecht, M. C. M. (2004). Particle-Based Multidimensional Multispecies Biofilm Model. Appl. Environ. Microbiol. 70, 5, 3024–3040.

[131] Picioreanu, C., van Loosdrecht, M. C. M., Heijnen, J. J. (1998). Mathematical modeling of biofilm structure with a hybrid differential-discrete cellular automaton approach. Biotechnol. Bioeng. 58, 1, 101–116.

[132] Picioreanu, C., van Loosdrecht, M. C. M., Heijnen, J. J. (2001). Two-dimensional model of biofilm detachment caused by internal stress from liquid flow. Biotechnol. Bioeng. 72, 2, 205–218.

[133] Poppele, E., Hozalski, R. M. (2003). Micro-cantilever method for measuring the tensile strength of biofilms and microbial flocs. J. Microbiol. Methods 55, 3, 607–615.

[134] Purevdorj-Gage, B. (2004). *Pseudomonas aeruginosa* biofilm structure, behavior and hydrodynamics. Master thesis, Montana State University-Billings.

[135] Purevdorj-Gage, J. W., B. Costerton, Stoodley, P. (2005). Phenotypic differentiation and seeding dispersal in non-mucoid and mucoid *Pseudomonas aeruginosa* biofilms. Microbiology 151, 1569–1576.

[136] Rehm, B. H., Valla, S. (1997). Bacterial alginates: biosynthesis and applications. Appl. Microbiol. Biotechnol. 48, 3, 281–288.

[137] Rice, S. A., Koh, K. S., Queck, S. Y., Labbate, M., Lam, K. W., Kjelleberg, S. (2005). Biofilm formation and sloughing in serratia marcescens are controlled by quorum sensing and nutrient cues. J. Bacteriol. 187, 10, 3477–3485.

[138] Rowley, J. A., Madlambayan, G., Mooney, D. J. (1999). Alginate hydrogels as synthetic extracellular matrix materials. Biomaterials 20, 1, 45–53.

[139] Sanderson, G. R., Ortega, D. (1993). Alginates and Gellan Gum: Complementary Gelling Agents. In: Food Hydrocolloids (Nishinari, K. , Doi, E. (eds.)). Springer, New York.

[140] Scheffler, C. (2009). Simulation gekoppelter Relaxations- und Erholungsprozesse bei technischen Gummiwerkstoffen mittels rheologischer Modelle. PhD thesis, Technische Universität Chemnitz.

[141] Shimazaki, T., Ogino, K. (1996). Viscoelastic properties of Ca^{2+}- and Na^+-gellan gum aqueous solutions. Carbohydr. Polym. 30, 2-3, 95–100.

[142] Shipley, R. J., Jones, G. W., Dyson, R. J., Sengers, B. G., Bailey, C. L., Catt, C. J., Please, C. P., Malda, J. (2009). Design criteria for a printed tissue engineering construct: a mathematical homogenization approach. J. Theor. Biol. 259, 3, 489–502.

[143] Shoichet, M. S., Li, R. H., White, M. L., Winn, S. R. (1996). Stability of hydrogels used in cell encapsulation: An in vitro comparison of alginate and agarose. Biotechnol. Bioeng. 50, 4, 374–381.

[144] Siebertz, K., van Bebber, D., Hochkirchen, T. (2010). Statistische Versuchsplanung. Springer, Berlin-Heidelberg.

[145] Stammen, S., Müller, B. K., Korneli, C., Biedendieck, R., Gamer, M., Franco-Lara, E., Jahn, D. (2010). High-yield intra- and extracellular protein production using *Bacillus megaterium*. Appl. Environ. Microbiol. 76, 12, 4037–4046.

[146] Staudt, C., Horn, H., Hempel, D. C., Neu, T. R. (2004). Volumetric measurements of bacterial cells and extracellular polymeric substance glycoconjugates in biofilms. Biotechnol. Bioeng. 88, 5, 585–592.

[147] Stevens, L., Calvert, P., Wallace, G. G., Panhuis, M. i. H. (2013). Ionic-covalent entanglement hydrogels from gellan gum, carrageenan and an epoxy-amine. Soft Matter 9, 11, 3009.

[148] Stoodley, P., Cargo, R., Rupp, C. J., Wilson, S., Klapper, I. (2002). Biofilm material properties as related to shear-induced deformation and detachment phenomena. J. Ind. Microbiol. Biotechnol. 29, 6, 361–367.

[149] Stoodley, P., Lewandowski, Z., Boyle, J. D., Lappin-Scott, H. M. (1998). Oscillation characteristics of biofilm streamers in turbulent flowing water as related to drag and pressure drop. Biotechnol. Bioeng. 57, 5, 536–544.

[150] Stoodley, P., Lewandowski, Z., Boyle, J. D., Lappin-Scott, H. M. (1999). Structural deformation of bacterial biofilms caused by short-term fluctuations in fluid shear: an in situ investigation of biofilm rheology. Biotechnol. Bioeng. 65, 1, 83–92.

[151] Stoodley, P., Sauer, K., Davies, D. G., Costerton, J. W. (2002). Biofilms as complex differentiated communities. Annu. Rev. Microbiol. 56, 187–209.

[152] Strathmann, M., Griebe, T., Flemming, H.-C. (2000). Artificial biofilm model - a useful tool for biofilm research. Appl. Microbiol. Biotechnol. 54, 2, 231–237.

[153] Sutherland, I. W. (1990). Biotechnology of microbial exopolysaccharides Vol. 9 of Cambridge Studies in Biotechnology. Cambridge University Press, Cambridge.

[154] Sutherland, I. W. (2001). Biofilm exopolysaccharides: a strong and sticky framework. Microbiology 147, 1, 3–9.

[155] Sutherland, I. W. (2001). The biofilm matrix - an immobilized but dynamic microbial environment. Trends Microbiol. 9, 5, 222–227.

[156] Sworn, G., Sanderson, G. R., Gibson, W. (1995). Gellan gum fluid gels. Food Hydrocoll. 9, 4, 265–271.

[157] Taherzadeh, D., Picioreanu, C., Küttler, U., Simone, A., Wall, W. a., Horn, H. (2010). Computational study of the drag and oscillatory movement of biofilm streamers in fast flows. Biotechnol. Bioeng. 105, 3, 600–610.

[158] Takahashi, R., Akutu, M., Kubota, K., Nakamura, K. (1999). Characterization of gellan gum in aqueous NaCl solution. In: Physical Chemistry and Industrial Application of Gellan Gum (Nishinari, K. (ed.)). Springer, Berlin-Heidelberg.

[159] Tang, J., Tungb, M. A., Zeng, Y. (1999). Characterization of Gellan Gels Using Stress Relaxation. J. Food Eng. 38, 1998, 279–295.

[160] te Nijenhuis, K. (1990). Viscoelastic properties of thermoreversible Gels. In: Physical Networks: Polymers and Gels (Burchard, W. , Ross-Murphy, S. B. (eds.)). Springer, Berlin-Heidelberg.

[161] te Nijenhuis, K. (1997). Thermoreversible Networks Viscoelastic Properties and Structure of Gels. Springer, Berlin-Heidelberg.

[162] te Nijenhuis, K., McKinley, G. H., Spiegelberg, S., Barnes, H. A., Aksel, N., Heymann, L., Odell, J. A. (2007). Non-newtonian flows. In: Springer Handbook of Experimental Fluid Mechanics (Tropea, C. , Yarin, A. L. , Foss, J. F. (eds.)). Springer, Berlin-Heidelberg.

[163] Telgmann, U., Horn, H., Morgenroth, E. (2004). Influence of growth history on sloughing and erosion from biofilms. Water Res. 38, 17, 3671–3684.

[164] Teratsubo, M., Tanaka, Y., Saeki, S. (2002). Measurement of stress and strain during tensile testing of gellan gum gels: effect of deformation speed. Carbohydr. Polym. 47, 1, 1–5.

[165] Tolker-Nielsen, T., Brinch, U. C., Ragas, P. C., Andersen, J. . B., Jacobsen, C. S., Molin, S. (2000). Development and Dynamics of Biofilms. J. Bacteriol. 182, 22, 6482–6489.

[166] Towler, B. W., Cunningham, A. B., Stoodley, P., Mckittrick, L. (2007). A model of fluid-biofilm interaction using a Burger material law. Biotechnol. Bioeng. 96, 2, 259–271.

[167] Towler, B. W., Rupp, C. J. J., Cunningham, A. L. B., Stoodley, P. (2003). Viscoelastic properties of a mixed culture biofilm from rheometer creep analysis. Biofouling 19, 5, 279–285.

[168] van Loosdrecht, M. C. M., Heijnen, J. J., Eberl, H., Kreft, J., Picioreanu, C. (2002). Mathematical modelling of biofilm structures. A. van. Leeuw. J. Microb. 81, 1, 245–256.

[169] Wagner, M., Manz, B., Volke, F., Neu, T. R., Horn, H. (2010). Online assessment of biofilm development, sloughing and forced detachment in tube reactor by means of magnetic resonance microscopy. Biotechnol. Bioeng. 107, 1, 172–181.

[170] Wanner, O., Gujer, W. (1986). A multispecies biofilm model. Biotechnol. Bioeng. 28, 3, 314–328.

[171] Ward, P. G., Goff, M., Donner, M., Kaminsky, W., O'Connor, K. E. (2006). A Two Step Chemo-biotechnological Conversion of Polystyrene to a Biodegradable Thermoplastic. Environ. Sci. Technol. 40, 7, 2433–2437.

[172] Wäsche, S. (2002). Einfluss der Wachstumsbedingungen auf Stoffübergang und Struktur von Biofilmsystemen. ibvt-Schriftenreihe (Hempel, D. C. (ed.)) Vol. 14. PhD thesis, Technische Universität Braunschweig, FIT-Verlag, Paderborn.

[173] Wäsche, S., Horn, H., Hempel, D. C., Stefan, W. (2002). Influence of growth conditions on biofilm development and mass transfer at the bulk / biofilm interface. Measurement 36, 19, 4775–4784.

[174] Watase, M., Nishinari, K. (1993). Rheology and DSC of Curdlan - DMSO - Water Systems. In: Food Hydrocolloids (Nishinari, K. , Doi, E. (eds.)). Springer, New York.

[175] Wee, S., Wilkinson, B. J. (1988). Increased outer membrane ornithine-containing lipid and lysozyme penetrability of *Paracoccus denitrificans* grown in a complex medium deficient in divalent cations. J. Bacteriol. 170, 3283–3286.

[176] Weller, D. M. (1988). Biological Control of Soilborne Plant Pathogens in the Rhizosphere with Bacteria. Annu. Rev. Phytopathol. 26, 1, 379–407.

[177] Wimpenny, J. W. T., Colasanti, R. (1997). A unifying hypothesis for the structure of microbial biofilms based on cellular automaton models. FEMS Microbiol. Ecol. 22, 1, 1–16.

[178] Wloka, M. (2006). Rheologische Untersuchungen an nativen Biofilmen von *Pseudomonas aeruginosa*. PhD thesis, Technische Universität Dortmund.

[179] Wloka, M., Rehage, H., Flemming, H.-C., Wingender, J. (2004). Rheological properties of viscoelastic biofilm extracellular polymeric substances and comparison to the behavior of calcium alginate gels. Colloid Polym. Sci. 282, 10, 1067–1076.

[180] Wloka, M., Rehage, H., Flemming, H.-C., Wingender, J. (2005). Structure and rheological behaviour of the extracellular polymeric substance network of mucoid *Pseudomonas aeruginosa* biofilms. Biofilms 2, 04, 275–283.

[181] Xavier, J. B., Picioreanu, C., Rani, S. A., van Loosdrecht, M. C. M., Stewart, P. S. (2005). Biofilm-control strategies based on enzymic disruption of the extracellular polymeric substance matrix - a modelling study. Microbiology 151, 3817–3832.

[182] Yamakawa, H. (1997). Models for polymer chains. In: Helical Wormlike Chains in Polymer Solutions. Springer, Berlin-Heidelberg.

[183] Yuguchi, Y. (1996). Small angle X-ray characterization of gellan gum containing a high content of sodium in aqueous solution. Carbohydr. Polym. 30, 2-3, 83–93.

[184] Yuguchi, Y., Urakawa, H., Kitamura, S., Wataoka, I., Kajiwara, K. (1999). The sol-gel transition of gellan gum aqueous solutions in the presence of various metal salts. In: Physical Chemistry and Industrial Application of Gellan Gum (Nishinari, K. (ed.)). Springer, Berlin-Heidelberg.

A Appendix

A.1 Modified BTR

A biofilm tube reactor (BTR) [61, 103, 173] was used to culture the biofilms. However, for the initial project it did not provide a sufficient connectivity for, e.g., microforce sensors, microelectrodes (pO_2 sensors, pH electrodes) or for a stereoscopic camera system. Taking pictures with good quality required a good contrast limiting the distance between camera objective and biofilm. Images were taken of the biofilm surfaces, thus were disturbed by the fluid passing by. To avoid a complete change of hydrodynamics in the proposed modified BTR the cross-sectional area was kept identical to the original design but changed from circular to elliptical. This made it possible to shrink one diameter by a factor of 2 to 13 mm. Fig. A.1 shows the proposed design. The planned BTR had several ports and a thermo regulation via cooler to allow a controlled gelation of a thermoreversible hydrogel functioning as biomimetic biofilm.

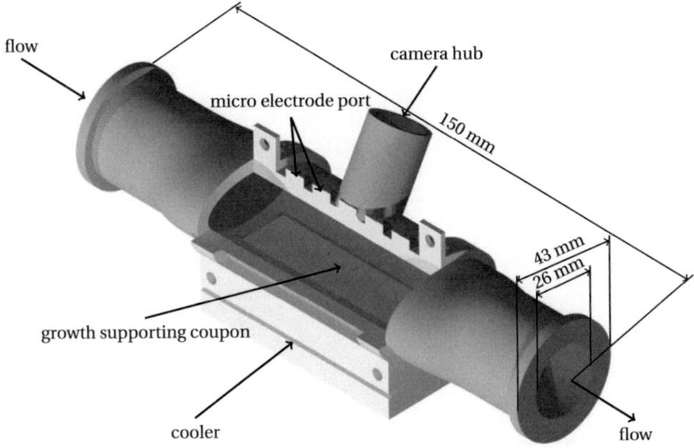

Figure A.1: Modified Biofilm Tube Reactor.

A.2 Parameters of Maxwell-Wiechert-Model

Table A.1: Parameters of Maxwell-Wiechert-model with four Maxwell elements for all 15 hydrogels.

Hydrogel #	Maxwell element i					
i/k	1	2	3	4	G', R^2	G'', R^2
τ_i [s]	0.077	1.062	23.136	1338		
$G_{i,1}$ [kPa]	0.377	0.249	0.510	2.894	0.949	0.653
$G_{i,2}$ [kPa]	0.541	0.362	0.549	3.813	0.945	0.674
$G_{i,3}$ [kPa]	0.423	0.242	0.525	2.114	0.955	0.794
$G_{i,4}$ [kPa]	1.551	1.103	2.300	6.329	0.949	0.753
$G_{i,5}$ [kPa]	1.297	0.768	1.395	7.240	0.948	0.640
$G_{i,6}$ [kPa]	2.957	2.101	3.740	14.132	0.955	0.505
$G_{i,7}$ [kPa]	1.296	0.767	1.639	7.192	0.910	0.773
$G_{i,8}$ [kPa]	1.275	0.737	1.189	7.874	0.923	0.751
$G_{i,9}$ [kPa]	0.298	0.271	0.389	1.845	0.905	0.162
$G_{i,10}$ [kPa]	2.198	1.502	2.275	7.798	0.951	0.611
$G_{i,11}$ [kPa]	2.147	1.428	3.085	3.583	0.782	0.674
$G_{i,12}$ [kPa]	0.790	0.506	0.819	4.210	0.935	0.771
$G_{i,13}$ [kPa]	0.318	0.280	0.231	0.986	0.960	0.013
$G_{i,14}$ [kPa]	1.752	1.503	2.176	8.180	0.971	0.065
$G_{i,15}$ [kPa]	2.400	1.316	1.703	9.864	0.872	0.760

Band 1 **Sunder, Matthias**: Oxidation grundwasserrelevanter Spurenverunreinigungen mit Ozon und Wasserstoffperoxid im Rohrreaktor. 1996. FIT-Verlag · Paderborn, ISBN 3-932252-00-4

Band 2 **Pack, Hubertus**: Schwermetalle in Abwasserströmen: Biosorption und Auswirkung auf eine schadstoffabbauende Bakterienkultur. 1996. FIT-Verlag · Paderborn, ISBN 3-932252-01-2

Band 3 **Brüggenthies, Antje**: Biologische Reinigung EDTA-haltiger Abwässer. 1996. FIT-Verlag · Paderborn, ISBN 3-932252-02-0

Band 4 **Liebelt, Uwe**: Anaerobe Teilstrombehandlung von Restflotten der Reaktivfärberei. 1997. FIT-Verlag · Paderborn, ISBN 3-932252-03-9

Band 5 **Mann, Volker G.**: Optimierung und Scale up eines Suspensionsreaktorverfahrens zur biologischen Reinigung feinkörniger, kontaminierter Böden. 1997. FIT-Verlag · Paderborn, ISBN 3-932252-04-7

Band 6 **Boll Marco**: Einsatz von Fuzzy-Control zur Regelung verfahrenstechnischer Prozesse. 1997. FIT-Verlag · Paderborn, ISBN 3-932252-06-3

Band 7 **Büscher, Klaus**: Bestimmung von mechanischen Beanspruchungen in Zweiphasenreaktoren. 1997. FIT-Verlag · Paderborn, ISBN 3-932252-07-1

Band 8 **Burghardt, Rudolf**: Alkalische Hydrolyse – Charakterisierung und Anwendung einer Aufschlußmethode für industrielle Belebtschlämme. 1998. FIT-Verlag · Paderborn, ISBN 3-932252-13-6

Band 9 **Hemmi, Martin**: Biologisch-chemische Behandlung von Färbereiabwässern in einem Sequencing Batch Process. 1999. FIT-Verlag · Paderborn, ISBN 3-932252-14-4

Schriftenreihe des Instituts für Bioverfahrenstechnik
der Technischen Universität Braunschweig

Band 10 Dziallas, Holger: Lokale Phasengehalte in zwei- und dreiphasig betriebenen Blasensäulenreaktoren. 2000. FIT-Verlag · Paderborn, ISBN 3-932252-15-2

Band 11 Scheminski, Anke: Teiloxidation von Faulschlämmen mit Ozon. 2001. FIT-Verlag · Paderborn, ISBN 3-932252-16-0

Band 12 Mahnke, Eike Ulf: Fluiddynamisch induzierte Partikelbeanspruchung in pneumatisch gerührten Mehrphasenreaktoren. 2002. FIT-Verlag · Paderborn, ISBN 3-932252-17-9

Band 13 Michele, Volker: CDF modeling and measurement of liquid flow structure and phase holdup in two- and three-phase bubble columns. 2002. FIT-Verlag · Paderborn, ISBN 3-932252-18-7

Band 14 Wäsche, Stefan: Einfluss der Wachstumsbedingungen auf Stoffübergang und Struktur von Biofilmsystemen. 2003. FIT-Verlag · Paderborn, ISBN 3-932252-19-5

Band 15 Krull Rainer: Produktionsintegrierte Behandlung industrieller Abwässer zur Schließung von Stoffkreisläuren. 2003. FIT-Verlag · Paderborn, ISBN 3-932252-20-9

Band 16 Otto, Peter: Entwicklung eines chemisch-biologischen Verfahrens zur Reinigung EDTA enthaltender Abwässer. 2003. FIT-Verlag · Paderborn, ISBN 3-932252-21-7

Band 17 Horn, Harald: Modellierung von Stoffumsatz und Stofftransport in Biofilmsystemen. 2003. FIT-Verlag · Paderborn, ISBN 3-932252-22-5

Band 18 Mora Naranjo, Nelson: Analyse und Modellierung anaerober Abbauprozesse in Deponien. 2004. FIT-Verlag · Paderborn, ISBN 3-932252-23-3

Band 19 Döpkens, Eckart: Abwasserbehandlung und Prozesswasserrecycling in der Textilindustrie. 2004. FIT-Verlag · Paderborn, ISBN 3-932252-24-1

Band 20 Haarstrick, Andreas: Modellierung millieugesteuerter biologischer Abbauprozesse in heterogenen problembelasteten Systemen. 2005. FIT-Verlag · Paderborn, ISBN 3-932252-27-6

Band 21 Baaß, Anne-Christina: Mikrobieller Abbau der Polyaminopolycarbonsäuren Propylendiamintetraacetat (PDTA) und Diethylentriaminpentaacetat (DTPA). 2004. FIT-Verlag · Paderborn, ISBN 3-932252-26-8

Band 22 Staudt, Christian: Entwicklung der Struktur von Biofilmen. 2006. FIT-Verlag · Paderborn, ISBN 3-932252-28-4

Band 23 Pilz, Roman Daniel: Partikelbeanspruchung in mehrphasig betriebenen Airlift-Reaktoren. 2006. FIT-Verlag · Paderborn, ISBN 3-932252-29-2

Band 24 Schallenberg, Jörg: Modellierung von zwei- und dreiphasigen Strömungen in Blasensäulenreaktoren. 2006. FIT-Verlag · Paderborn, ISBN 3-932252-30-6

Band 25 Enß, Jan Hendrik: Einfluss der Viskosität auf Blasensäulenströmungen. 2006. FIT-Verlag · Paderborn, ISBN 3-932252-31-4

Band 26 Kelly, Sven: Fluiddynamischer Einfluss auf die Morphogenese von Biopellets filamentöser Pilze. 2006. FIT-Verlag · Paderborn, ISBN 3-932252-32-2

Band 27 Grimm, Luis Hermann: Sporenaggregationsmodell für die submerse Kultivierung koagulativer Myzelbildner. 2006. FIT-Verlag · Paderborn, ISBN 3-932252-33-0

Band 28 León Ohl, Andrés: Wechselwirkungen von Stofftransport und Wachstum in Biofilsystemen. 2007. FIT-Verlag · Paderborn, ISBN 3-932252-34-9

Band 29 Emmler, Markus: Freisetzung von Glucoamylase in Kultivierungen mit *Aspergillus niger*. 2007. FIT-Verlag · Paderborn, ISBN 3-932252-35-7

Band 30 Leonhäuser, Johannes: Biotechnologische Verfahren zur Reinigung von quecksilberhaltigem Abwasser. 2007. FIT-Verlag · Paderborn, ISBN 3-932252-36-5

Band 31 Jungebloud, Anke: Untersuchung der Genexpression in *Aspergillus niger* mittels Echtzeit-PCR. 1996. FIT-Verlag · Paderborn, ISBN 978-3-932252-37-2

Band 32 Hille, Andrea: Stofftransport und Stoffumsatz in filamentösen Pilzpellets. 2008. FIT-Verlag · Paderborn, ISBN 978-3-932252-38-9

Band 33 Fürch, Tobias: Metabolic characterization of recombinant protein production in *Bacillus megaterium*. 2008. FIT-Verlag · Paderborn, ISBN 978-3-932252-39-6

Band 34 Grote, Andreas Georg: Datenbanksysteme und bioinformatische Werkzeuge zur Optimierung biotechnologischer Prozesse mit Pilzen. 2008. FIT-Verlag · Paderborn, ISBN 978-3-932252-40-120

Band 35 Möhle, Roland Bernhard: An Analytic-Synthetic Approach Combining Mathematical Modeling and Experiments – Towards an Understanding of Biofilm Systems. 2008. FIT-Verlag · Paderborn, ISBN 978-3-932252-41-9

Band 36 Reichel, Thomas: Modelle für die Beschreibung das Emissionsverhaltens von Siedlungsabfällen. 2008. FIT-Verlag · Paderborn, ISBN 978-3-932252-42-6

Band 37 Schultheiss, Ellen: Charakterisierung des Exopolysaccharids PS-EDIV von *Sphingomonas pituitosa*. 2008. FIT-Verlag · Paderborn, ISBN 978-3-932252-43-3

Band 38 Dreger, Michael Andreas: Produktion und Aufarbeitung des Exopolysaccharids PS-EDIV aus *Sphingomonas pituitosa*. 1996. FIT-Verlag · Paderborn, ISBN 978-3-932252-44-0

Band 39 Wiebels, Cornelia: A Novel Bubble Size Measuring Technique for High Bubble Density Flows. 2009. FIT-Verlag · Paderborn, ISBN 978-3-932252-45-7

Band 40 Bohle, Kathrin: Morphologie- und produktionsrelevante Gen- und Proteinexpression in submersen Kultivierungen von *Aspergillus niger*. 2009. FIT-Verlag · Paderborn, ISBN 978-3-932252-46-2

Band 41 Fallet, Claas: Reaktionstechnische Untersuchungen der mikrobiellen Stressantwort und ihrer biotechnologischen Anwendungen. 2009. FIT-Verlag · Paderborn, ISBN 978-3-932252-47-1

Band 42 Vetter, Andreas: Sequential Co-simulation as Method to Couple CFD and Biological Growth in a Yeast. 2009. FIT-Verlag · Paderborn, ISBN 978-3-932252-48-8

Schriftenreihe des Instituts für Bioverfahrenstechnik
der Technischen Universität Braunschweig

Band 43 Jung, Thomas: Einsatz chemischer Oxidationsverfahren zur Behandlung industrieller Abwässer. 2010. FIT-Verlag · Paderborn, ISBN 978-3-932252-49-5

Band 45 Herrmann, Tim: Transport von Proteinen in Partikeln der Hydrophoben Interaktions Chromatographie. 2010. FIT-Verlag · Paderborn, ISBN 978-3-932252-51-8

Band 46 Becker, Judith: Systems Metabolic Engineering of *Corynebacterium glutamicum* towards improved Lysine Prodction. 2010. Cuvillier-Verlage · Göttingen, ISBN 978-3-86955-426-6

Band 47 Melzer, Guido: Metabolic Network Analysis of the Cell Factory *Aspergillus niger*. 2010. Cuvillier-Verlage · Göttingen, ISBN 978-3-86955-456-3

Band 48 Bolten J., Christoph: Bio-based Production of L-Methionine in *Corynebacterium glutamicum*. 2010. Cuvillier-Verlage · Göttingen, ISBN 978-3-86955-486-0

Band 49 Lüders, Svenja: Prozess- und Proteomanalyse gestresster Mikroorganismen. 2010. Cuvillier-Verlage · Göttingen, ISBN 978-3-86955-435-8

Band 50 Wittmann, Christoph: Entwicklung und Einsatz neuer Tools zur metabolischen Netzwerkanalyse des industriellen Aminosäure-Produzenten *Corynebacterium glutamicum*. 2010. Cuvillier-Verlage · Göttingen, ISBN 978-3-86955-445-7

Band 51 Edlich, Astrid: Entwicklung eines Mikroreaktors als Screening-Instrument für biologische Prozesse. 2010. Cuvillier-Verlage · Göttingen, ISBN 978-3-86955-470-9

Band 52 Hage, Kerstin: Bioprozessoptimierung und Metabolomanalyse zur Proteinproduktion in *Bacillus licheniformis*. 2010. Cuvillier-Verlage · Göttingen, ISBN 978-3-86955-578-2

Band 53 Kiep, Katina Andrea: Einfluss von Kultivierungsparametern auf die Morphologie und Produktbildung von *Aspergillus niger*. 2010. Cuvillier-Verlage · Göttingen, ISBN 978-3-86955-632-1

Band 54 Fischer, Nicole: Experimental investigations on the influence of physico-chemical parameters on anaerobic degradation in MBT residual waste. 2011. Cuvillier-Verlage · Göttingen, ISBN 978-3-86955-679-6

Band 55 Schädel, Friederike: Stressantwort von Mikroorganismen. 2011. Cuvillier-Verlage · Göttingen, ISBN 978-3-86955-746-5

Band 56 Wichter, Johannes: Untersuchung der L-Cystein-Biosynthese in *Escherichia coli* mit Techniken der Metabolom- und ^{13}C-Stoffflussanalyse. 2011. Cuvillier-Verlage · Göttingen, ISBN 978-3-86955-750-2

Band 57 Knappik, Irena Isabell: Charakterisierung der biologischen und chemischen Reaktionsprozesse in Siedlungsabfällen. 2011. Cuvillier-Verlage · Göttingen, ISBN 978-3-86955-760-1

Band 58 Driouch, Habib: Systems biotechnology of recombinant protein production in *Aspergillus niger*. 2011. Cuvillier-Verlag Göttingen, ISBN 978-3-86955-808-0

Band 59 Gehder, Matthias: Development and Validation of Indicators for the Production and Quality of Seed Cultures. 2011. Cuvillier-Verlag Göttingen, ISBN 978-3-86955-847-9

Schriftenreihe des Instituts für Bioverfahrenstechnik
der Technischen Universität Braunschweig

Band 60 Sommer, Becky: Methodenentwicklung zur Charakterisierung sporenbildender Pilz-Seedingkulturen. 2011. Cuvillier-Verlag Göttingen, ISBN 978-3-86955-851-6

Band 61 Dohnt, Katrin: Charakterisierung von *Pseudomonas aeruginosa*-Biofilmen in einem *in vitro*-Harnwegskathetersystem. 2011. Cuvillier-Verlag Göttingen, ISBN 978-3-86955-852-3

Band 62 Greis, Tillman: Meddling the risk of chlorinated hydrocarbons in urban groundwater. 2011. Cuvillier-Verlag Göttingen, ISBN 978-3-86955-970-4

Band 63 David, Florian: Holistic bioprocess engineering of antibody fragment secreting *Bacillus megaterium*. 2012. Cuvillier-Verlag Göttingen, ISBN 978-3-95404-115-2

Band 64 Palme, Wiebke: Taxonomische Einordnung des Polyaminopolycarbonsäure-abbauenden Stammes BNC1 und Untersuchungen zum Abbau von 1,3 Propylendiamintetraacetat. 2012. Cuvillier-Verlag Göttingen, ISBN 978-3-95404-158-9

Band 65 Lin, Pey-Jin: Effect of fluid dynamics on pellet morphology and product formation of *Aspergillus niger*. 2012. Cuvillier-Verlag Göttingen, ISBN 978-3-95404-181-7

Band 66 Kind, Stefanie: Synthetic Metabolic Engineering of *Corynebacterium glutamicum* for Bio-based Production of 1,5-Diaminopentane. 2012. Cuvillier-Verlag Göttingen, ISBN 978-3-95404-264-7

Band 67 Wilk, Franziska: Charakterisierung der Stoffströme vorbehandelter Siedlungsabfälle in Deponiebioreaktoren. 2012. Cuvillier-Verlag Göttingen, ISBN 978-3-95404-281-4

Band 68 Korneli, Claudia: Target-oriented Bioprocess Optimization for Recombinant Protein Production in *Bacillus megaterium*. 2012. Cuvillier-Verlag Göttingen, ISBN 978-3-95404-289-0

Band 69 Eslahpazir, Manely: Numerical Characterization of Mechanical Stress and Flow Patterns in Stirred Tank Bioreactors. 2013. Cuvillier-Verlag Göttingen, ISBN 978-3-95404-449-8

Band 70 Wucherpfennig, T.: Cellular Morphology – A novel Process Parameter for the Cultivation of Eukaryotic Cells. 2013. Cuvillier-Verlag Göttingen, ISBN 978-3-95404-456-6

Band 71 Buschke, Nele: Bio-Nylon Monomers from Renewables using *Corynebacterium glutamicum*. 2013. Cuvillier-Verlag Göttingen, ISBN 978-3-95404-457-3

Band 72 Bergmann, Sven: Ectoine production by halotolerant microorganisms. 2013. Cuvillier-Verlag Göttingen, ISBN 978-3-95404-556-3